The IMA Volumes in Mathematics and its Applications

Volume 86

Series Editors
Avner Friedman   Robert Gulliver

Springer
New York
Berlin
Heidelberg
Barcelona
Budapest
Hong Kong
London
Milan
Paris
Santa Clara
Singapore
Tokyo

Institute for Mathematics and its Applications
IMA

The **Institute for Mathematics and its Applications** was established by a grant from the National Science Foundation to the University of Minnesota in 1982. The IMA seeks to encourage the development and study of fresh mathematical concepts and questions of concern to the other sciences by bringing together mathematicians and scientists from diverse fields in an atmosphere that will stimulate discussion and collaboration.

The IMA Volumes are intended to involve the broader scientific community in this process.

Avner Friedman, Director

Robert Gulliver, Associate Director

* * * * * * * * * *

IMA ANNUAL PROGRAMS

1982–1983	Statistical and Continuum Approaches to Phase Transition
1983–1984	Mathematical Models for the Economics of Decentralized Resource Allocation
1984–1985	Continuum Physics and Partial Differential Equations
1985–1986	Stochastic Differential Equations and Their Applications
1986–1987	Scientific Computation
1987–1988	Applied Combinatorics
1988–1989	Nonlinear Waves
1989–1990	Dynamical Systems and Their Applications
1990–1991	Phase Transitions and Free Boundaries
1991–1992	Applied Linear Algebra
1992–1993	Control Theory and its Applications
1993–1994	Emerging Applications of Probability
1994–1995	Waves and Scattering
1995–1996	Mathematical Methods in Material Science
1996–1997	Mathematics of High Performance Computing
1997–1998	Emerging Applications of Dynamical Systems
1998–1999	Mathematics in Biology

Continued at the back

Bjorn Engquist Gregory A. Kriegsmann
Editors

Computational Wave Propagation

With 69 Illustrations

Springer

Bjorn Engquist
Department of Mathematics
University of California, Los Angeles
405 Hilgard Ave.
Los Angeles, CA 90024 USA

Gregory A. Kriegsmann
Department of Mathematics
Center for Applied Mathematics and Statistics
New Jersey Institute of Technology
University Heights
Newark, NJ 07102 USA

Series Editors:
Avner Friedman
Robert Gulliver
Institute for Mathematics and its Applications
University of Minnesota
Minneapolis, MN 55455 USA

Mathematics Subject Classifications (1991): 65M, 65N, 78A25, 78A45, 78A50, 86A15

Library of Congress Cataloging-in-Publication Data
Computational wave propagation / editors, Bjorn Engquist and Gregory A. Kriegsmann.
p. cm. – (The IMA volumes in mathematics and its applications ; v. 86)
Includes bibliographical references and index.
ISBN-13: 978-1-4612-7531-2 e-ISBN-13: 978-1-4612-2422-8
DOI: 10.1007/978-1-4612-2422-8
1. Electromagnetic waves – Mathematics. 2. Electromagnetic waves – Transmission. 3. Electromagnetic waves – Scattering. I. Engquist, Bjorn, 1945- . II. Kriegsmann, Gregory A. III. Series.
QC661.C654 1997
531′.1133 – dc20 96-38282

Printed on acid-free paper.

Softcover reprint of the hardcover 1st edition 1997

Production managed by Allan Abrams; manufacturing supervised by Jacqui Ashri.
Camera-ready copy prepared by the IMA.

9 8 7 6 5 4 3 2 1

FOREWORD

This IMA Volume in Mathematics and its Applications

COMPUTATIONAL WAVE PROPAGATION

is based on the workshop with the same title and was an integral part of the 1994–1995 IMA program on "Waves and Scattering." We would like to thank Bjorn Engquist and Gregory A. Kriegsmann for their hard work in organizing this meeting and in editing the proceedings. We also take this opportunity to thank the National Science Foundation, the Army Research Office, and the Office of Naval Research, whose financial support made this workshop possible.

Avner Friedman

Robert Gulliver

PREFACE

Although the field of wave propagation and scattering has its classical roots in the last century, it has enjoyed a rich and vibrant life over the past 50 odd years. Scientists, engineers, and mathematicians have developed sophisticated asymptotic and numerical tools to solve problems of ever increasing complexity. Their work has been spurred on by emerging and maturing technologies, primarily concerned with the propagation and reception of information, and the efficient transmission of energy.

The vitality of this scientific field is not waning. Increased demands to precisely quantify, measure, and control the propagation and scattering of waves in increasingly complex settings pose challenging scientific and mathematical problems. These push the envelope of analysis and computing, just as their forerunners did 50 years ago. These modern technological problems range from using underwater sound to monitor and predict global warming, to periodically embedding phase-sensitive amplifiers in optical fibers to insure long range digital communication.

The papers contained in this volume represent a snap-shot of current applied mathematical research in wave propagation and scattering. Although the mathematical underpinnings of the research contained herein are rooted in classical asymptotic and numerical analyses, each author is motivated by a complex technological problem which requires a resolution. It is our hope that the efforts of these authors will intrigue and stimulate the reader to pursue research into this rich and diverse scientific field.

Bjorn Engquist

Gregory A. Kriegsmann

CONTENTS

CONTENTS

ON HIGH-ORDER RADIATION BOUNDARY CONDITIONS

THOMAS HAGSTROM*

Abstract. In this paper we develop the theory of high-order radiation boundary conditions for wave propagation problems. In particular, we study the convergence of sequences of time-local approximate conditions to the exact boundary condition, and subsequently estimate the error in the solutions obtained using these approximations. We show that for finite times the Padé approximants proposed by Engquist and Majda lead to exponential convergence if the solution is smooth, but that good long-time error estimates cannot hold for spatially local conditions. Applications in fluid dynamics are also discussed.

Key words. Radiation boundary conditions, integral equations, hyperbolic systems.

1. Introduction. Problems in wave propagation are generally posed on unbounded domains. Their numerical solution thus requires the introduction of an artificial boundary and the imposition of radiation boundary conditions there. Scores of authors have considered this problem, and a number of reasonably accurate procedures have been discovered. Nonetheless, in order to obtain some specified accuracy, it is still generally the practice to enlarge the domain - a process which may be inefficient and difficult to automate.

In this work we pursue a different approach - namely to fix the artificial boundary and to improve the accuracy by increasing the order of the approximate radiation conditions. From a practical point of view, we see that these high-order conditions can be easily implemented via the introduction of auxiliary functions on the boundary. From a theoretical point of view, estimates of convergence for fixed boundaries and increasing order are needed. We develop such estimates for the wave equation in a half-space by first finding a convenient representation of the exact radiation condition, which turns out to involve convolution in time with a Bessel kernel. Approximate conditions are similarly represented in terms of convolutions, and the error then depends on the difference between the exact and approximate kernels. Using approximations to an integral representation of the exact kernel, convergent time-local approximate conditions are derived. These include the spatially local Padé conditions proposed by Engquist and Majda [7,8]. For long time computations, on the other hand, it is shown that spatially nonlocal conditions are generally needed.

Generalizations to other problems are presented, including the linearized Euler equations as well as the wave equation with circular and

* Department of Mathematics and Statistics, The University of New Mexico, Albuquerque, NM 87131. Supported, in part, by the Institute for Computational Mechanics in Propulsion (ICOMP), NASA, Lewis Research Center, Cleveland, Ohio, and by NSF Grant No. DMS-9304406. Part of the work was also carried out during a visit to the IMA.

spherical boundaries. Throughout we indicate some interesting theoretical and practical issues which remain unresolved.

2. The wave equation in a half-space.

2.1. Exact boundary conditions. We consider:

$$u_{tt} = c^2\nabla^2 u + f, \quad t > 0, \quad x = (x_1, y) \in (0,\infty) \times R^{n-1}, \tag{2.1}$$

$$u(x,0) = g(x), \quad B_0 u(0,y,t) = g_0(y,t). \tag{2.2}$$

We suppose, for some $L, \delta > 0$, that $f = g = 0$ and $c = 1$ for $x_1 \geq L - \delta$. Let $\hat{u}(x_1, k, s)$ be the Fourier-Laplace transform of u with respect to y and t. Then it is easily shown (e.g. [11]) that $\hat{u}$ satisfies the exact boundary condition at $x_1 = L$:

$$\frac{\partial \hat{u}}{\partial x_1} + \left(s^2 + |k|^2\right)^{1/2} \hat{u} = 0. \tag{2.3}$$

(The branch of $(s^2 + |k|^2)^{1/2}$ is chosen so that it is analytic in the right half s-plane and has positive real part.) This exact condition is expressed in terms of u in the following way: Let $\mathcal{F}$ denote Fourier transformation with respect to y and $\mathcal{F}^{-1}$ be its inverse. Let:

$$\mathcal{K}(t) = \frac{J_1(t)}{t} = \frac{1}{\pi}\int_{-1}^{1} \sqrt{1-w^2}\cos wt\,dw. \tag{2.4}$$

As shown in the appendix,

$$\hat{\mathcal{K}}(s) = \left(s^2+1\right)^{1/2} - s. \tag{2.5}$$

Using standard formulas from Laplace transform theory (e.g. [5]) we finally have the exact condition at $x_1 = L$:

$$\frac{\partial u}{\partial x_1} + \frac{\partial u}{\partial t} + \mathcal{F}^{-1}\left(|k|^2\mathcal{K}(|k|t) * (\mathcal{F}u(x_1,\cdot,t))\right) = 0. \tag{2.6}$$

(Here, $*$ denotes convolution.)

2.2. Approximate conditions. Although it may be possible to directly implement (2.6) using FFT's and fast convolutions, most work has been focussed on the development of approximate conditions involving differential operators in time and, usually, space. Local approximations in time correspond to rational approximations in s to $\hat{\mathcal{K}}$:

$$\left(s^2+|k|^2\right)^{1/2} - s = |k|\left(\left(z^2+1\right)^{1/2} - z\right) \approx |k|R(z), \quad z = s/|k|. \tag{2.7}$$

We take R to be a rational function of degree $(p, p+1)$, that is,

$$R(z) = \frac{P(z)}{Q(z)}, \quad \deg(P) = p, \quad \deg(Q) = p+1. \tag{2.8}$$

The approximate condition may be directly localized in time by applying the operator $\mathcal{F}^{-1}Q(|k|^{-1}\partial/\partial t)\mathcal{F}$. However, this leads to differential operators of high order as p is increased. To develop a more convenient framework for implementation, we make the additional assumption that the roots of Q are distinct. Then, R has a partial fraction expansion:

$$R(z) = \sum_{j=1}^{p+1} \frac{\alpha_j}{z-\rho_j}. \tag{2.9}$$

Let

$$\hat{h}_j = \frac{\alpha_j |k|}{z-\rho_j}\hat{u}, \tag{2.10}$$

and let $\bar{v}(x_1, |k|, t)$ be the Fourier transform of v with respect to y. Here, v denotes the approximation to u computed on the bounded domain. We finally have the approximate boundary condition:

$$\frac{\partial \bar{v}}{\partial x_1} + \frac{\partial \bar{v}}{\partial t} + \sum_{j=1}^{p+1} \bar{h}_j = 0, \tag{2.11}$$

$$\left(\frac{\partial}{\partial t} - |k|\rho_j\right)\bar{h}_j = \alpha_j |k|^2 \bar{v}. \tag{2.12}$$

The advantage of this formulation is clear: the order is increased simply by increasing the number of terms in the sum. From the point of view of code development, this is very convenient.

We note that the conditions above are still nonlocal in space. For periodic problems this is no obstacle, as FFT's can be used. However, the nonlocality does preclude their use at more general boundaries. A glance at (2.12) reveals the condition for spatial locality: the poles, ρ_j, of $R(z)$ must come in conjugate, imaginary pairs, or be 0 and R itself must be an odd function of z. We may then assume that R has an expansion of the form:

$$R(z) = \sum_{j=1}^{q} \frac{\gamma_j z}{z^2+\beta_j^2}. \tag{2.13}$$

This leads to the local implementation:

$$\frac{\partial v}{\partial x_1} + \frac{\partial v}{\partial t} + \sum_{j=1}^{q} \phi_j = 0, \tag{2.14}$$

$$\left(\frac{\partial^2}{\partial t^2} - \beta_j^2 \nabla_{\tan}^2\right)\phi_j = -\gamma_j \nabla_{\tan}^2 \frac{\partial v}{\partial t}. \tag{2.15}$$

We would like to impose such a locality condition on $R(z)$, but it will be shown later that such approximations cannot lead to good error estimates uniformly in time or in tangential wave number.

In what follows, we will view

$$\mathcal{G}(t) = \mathcal{L}^{-1}R(s), \tag{2.16}$$

as an approximation to $\mathcal{K}(t)$. For reference we note that (2.9) corresponds to:

$$\mathcal{G}(t) = \sum_{j=1}^{p+1} \alpha_j e^{\rho_j t}, \tag{2.17}$$

while (2.13) implies:

$$\mathcal{G}(t) = \sum_{j=1}^{q} \gamma_j \cos\beta_j t. \tag{2.18}$$

2.3. Error estimates. We have seen that it is relatively straightforward to implement conditions of increasing order, at least in the half-space (or periodic) case. This leads to the question of convergence. Naturally, error estimates for approximate boundary conditions have been considered (e.g. [2],[8], [13],[20]). However, none of these consider convergence for a fixed problem in a fixed domain as the order of the conditions is increased.

Let $e = u - v$ be the error. Then e satisfies:

$$e_{tt} = c^2\nabla^2 e, \quad t > 0, \quad x = (x_1, y) \in (0, L) \times R^{n-1} \equiv \Omega, \tag{2.19}$$

$$e(x, 0) = 0, \quad B_0 e(0, y, t) = 0, \tag{2.20}$$

$$\frac{\partial \bar{e}}{\partial x_1} + \frac{\partial \bar{e}}{\partial t} + |k|^2\mathcal{G}(|k|t) * \bar{e} = |k|^2\mathcal{E}(|k|t) * \bar{u}. \tag{2.21}$$

The error kernel is given by:

$$\mathcal{E}(\tau) = \mathcal{G}(\tau) - \frac{J_1(\tau)}{\tau}. \tag{2.22}$$

Estimates of e naturally require both the stability and consistency of the approximate boundary conditions. Stability is a consequence of the uniform Lopatinski condition:

$$s + (s^2 + |k|^2)^{1/2} + |k|R(z) \neq 0, \tag{2.23}$$

for

$$\Re(s) \geq 0, \quad k \in R^{n-1}, \quad (s, k) \neq (0, 0). \tag{2.24}$$

Then we have (e.g. Sakamoto [22, Ch. 3]):

$$\int_0^T \|e(\cdot,t)\|^2_{H_1[\Omega]}dt \leq C^2 \int_0^T \|Eu(L,\cdot,t)\|^2_{H_0[R^{n-1}]}dt, \tag{2.25}$$

where,

$$Ew = \mathcal{F}^{-1}|k|^2\mathcal{E}(|k|t) * (\mathcal{F}w). \tag{2.26}$$

The error may now be bounded in terms of the error in the approximation, $\mathcal{G}$, to $\mathcal{K}$. In particular, suppose, for some $\mu \geq 0$ and $T \geq 1$:

$$\|\mathcal{E}\|_{L_1[(0,T)]} \leq \epsilon T^{\mu}. \tag{2.27}$$

Then, by Parseval's identity and standard estimates for convolutions,

$$\begin{aligned}\int_0^T \|Eu(L,\cdot,t)\|^2_{H_0[R^{n-1}]}dt &= \int_0^T \int_{R^{n-1}} |k|^4|(\mathcal{E}(|k|t) * \bar{u}(L,k,t))(t)dkdt \\ &\leq \int_0^T \int_{R^{n-1}} |k|^2|\bar{u}(L,k,t)|^2\|\mathcal{E}\|^2_{L_1[(0,|k|T)]} \\ &\leq \epsilon^2 T^{2\mu} \int_0^T \|u(L,\cdot,t)\|^2_{H_{1+\mu}[R^{n-1}]}dt.\end{aligned} \tag{2.28}$$

Substituting this into (2.25) we finally obtain:

$$\int_0^T \|e(\cdot,t)\|^2_{H_1[\Omega]}dt \leq \epsilon^2 T^{2\mu} C^2 \int_0^T \|u(L,\cdot,t)\|^2_{H_{1+\mu}[R^{n-1}]}dt. \tag{2.29}$$

This error estimate is best, both from the point of view of long time behavior and from the point of view of smoothness required of u, if $\mu = 0$. We note that such an estimate requires bounds on $\|\mathcal{E}\|_{L_1[(0,\infty)]}$. This cannot be attained for local conditions, as we have seen that they involve convolution kernels which are combinations of $\cos\beta_j t$ (2.18), and, hence, are not elements of $L_1[(0,\infty)]$. Time uniform estimates could be obtained using spatially nonlocal conditions, however. Some discussion of long-time behavior of spatially nonlocal boundary conditions appears in [6,17]. In [15] we construct conditions using Laguerre and exponential expansions. Although the conditions so derived do lead to estimates with $\mu = 0$, convergence as the order of approximation was increased was slow at best. Below we introduce a new nonlocal approximation based on the direct approximation to an integral representation for $\mathcal{K}(t)$.

Our point of view leads to an interesting, and to our knowledge unsolved, problem in approximation theory. Define $\mathcal{H}_p$ to be the set of all real functions in $L_1[(0,\infty)]$ whose Laplace transform is a rational function of degree $(p, p+1)$. More directly, a function in $\mathcal{H}_p$ takes the form:

$$\mathcal{G}(t) = \sum_{j=1}^{p_1} M_j(t)e^{-\gamma_j t}\cos(\beta_j t + \phi_j) + \sum_{j=1}^{p_2} N_j(t)e^{-\kappa_j t}, \tag{2.30}$$

where γ_j and κ_j are positive and $M_j(t)$ and $N_j(t)$ are polynomials. Here p is given by:

$$p+1 = 2\sum_{j=1}^{p_1}(\deg(M_j)+1) + \sum_{j=1}^{p_2}(\deg(N_j)+1). \tag{2.31}$$

Problem A: For fixed p characterize and find an algorithm to compute the best $L_1[(0,\infty)]$ approximation, $\mathcal{G} \in \mathcal{H}_p$, to $\mathcal{K}$ or to other kernels. Estimate the behavior of the error as p is increased.

The solution of this problem would provide us with optimal approximations to convolutions via the solution of differential equations. We note that for bounded intervals and sums excluding trigonometric terms (i.e. $p_1 = 0$), a theory does exist. (See Braess [3, Ch. 6].)

2.4. Methods derived via quadrature. In general, approximate boundary conditions have been derived either by direct approximation to the symbol (e.g. [24]) or through the use of far-field asymptotics (e.g. [2,16]). Here we show how a class of convergent local (in space and time) approximate conditions may be derived by approximating the integral representation of the exact kernel:

$$\frac{J_1(t)}{t} = \frac{1}{\pi}\int_{-1}^{1}\sqrt{1-w^2}\cos wt\,dw. \tag{2.32}$$

The simplest example is the trapezoid rule:

$$\frac{J_1(t)}{t} \approx \frac{2}{(q+1)\pi}\sum_{j=1}^{q}\sqrt{1-w_j^2}\cos w_j t, \quad w_j = -1 + \frac{2j}{q+1}. \tag{2.33}$$

After Laplace transformation we find that:

$$R(z) = \frac{2}{(q+1)\pi}\sum_{j=1}^{q}\frac{\sqrt{1-w_j^2}\,z}{z^2+w_j^2}. \tag{2.34}$$

(Of course, in implementations of the condition the number of terms in the sum can be halved using the evenness in w_j of the integrand.)

The well-posedness of these approximations follows directly from a check on the Lopatinski condition. (It also follows from general results for local conditions in [12,23].) In order to derive error estimates, however, we must explicitly bound the stability constant, C in (2.29), as a function of q. Using Parseval's relation (e.g. [21]) we have:

$$C \leq c_0(T) \sup_{s=\gamma+i\eta,\ k\in R^{n-1}} \left| \frac{(s^2+|k|^2)^{1/2}+|k|}{s+(s^2+|k|^2)^{1/2}+|k|R(s/|k|)} \right|. \tag{2.35}$$

Here $\gamma > 0$ is fixed and $c_0(T)$ depends on γ but is independent of the boundary condition. Dividing through by $|k|$ we see that our problem is reduced to the estimation of:

$$\sup_{\Re(z)\geq 0} |Q(z)|, \quad Q(z) = \frac{(z^2+1)^{1/2}+1}{z+(z^2+1)^{1/2}+R(z)}. \tag{2.36}$$

Noting the form of $R(z)$, and recalling the choice of branch for the roots, we see that the Q is bounded as $z \to \infty$ independent of q. Therefore, by the maximum principle, we can restrict attention to the imaginary axis, $z = i\eta$. In this case the poles of R are located on the imaginary axis with $|\eta| < 1$ and $|R|$ is strictly decreasing in $|\eta|$ for $|\eta| \geq 1$. Therefore, for $|\eta| \geq 1$ we have:

$$|Q| \leq \frac{1+\sqrt{\eta^2-1}}{|\eta|+\sqrt{\eta^2-1}-|R(i)|} \leq \frac{1}{1-|R(i)|}. \tag{2.37}$$

Now

$$1-|R(i)| = 1 - \frac{2}{(q+1)\pi}\sum_{j=1}^{q}\frac{1}{\sqrt{1-w_j^2}}. \tag{2.38}$$

To bound this we note:

$$\pi = \int_{-1}^{1}\frac{dw}{\sqrt{1-w^2}}. \tag{2.39}$$

For $q = 2p$, even, we have:

$$\pi > 2\sum_{j=1}^{p}\int_{w_j-2/(q+1)}^{w_j}\frac{dw}{\sqrt{1-w^2}}. \tag{2.40}$$

Generally we have:

$$\int_{w_j-2/(q+1)}^{w_j}\frac{dw}{\sqrt{1-w^2}} > \frac{2}{q+1}\frac{1}{\sqrt{1-w_j^2}}, \tag{2.41}$$

and for $j = 1$, $q \geq 2$;

$$\int_{-1}^{w_1}\frac{dw}{\sqrt{1-w^2}} > \frac{\sqrt{2}-\sqrt{3/2}}{\sqrt{q+1}} + \frac{2}{q+1}\frac{1}{\sqrt{1-w_1^2}}. \tag{2.42}$$

Hence, for some constant, c, independent of q:

$$\frac{1}{1-|R(i)|} \leq c\sqrt{q}. \tag{2.43}$$

For $q = 2p+1$, odd,

$$\pi = 2\sum_{j=1}^{p+1} \int_{w_j-2/(q+1)}^{w_j} \frac{dw}{\sqrt{1-w^2}}, \tag{2.44}$$

and (2.43) is similarly established. For $|\eta| \leq 1$ we distinguish two cases, $|\eta| < |w_1|$ and $|\eta| > |w_1|$. For the former we have:

$$|\mathcal{Q}| \leq \frac{2}{\sqrt{1-w_1^2}} \leq c\sqrt{q}, \tag{2.45}$$

while for the latter it is easily shown that $|\mathcal{Q}|$ is maximized at $|\eta| = 1$, which we have estimated above. Our final estimate is:

$$C \leq \bar{C}(T)\sqrt{q}, \tag{2.46}$$

with $\bar{C}$ independent of q.

To derive error estimates we first estimate the error in the quadrature formula for fixed t. Although the integrand doesn't possess a bounded derivative, its derivative is integrable. Therefore, employing the Peano Kernel representation of the error:

$$\left|\frac{J_1(t)}{t} - \frac{2}{(q+1)\pi}\sum_{j=1}^{q}\sqrt{1-w_j^2}\cos w_j t\right| \leq \frac{c_1 + 2c_2 t}{q}. \tag{2.47}$$

Here, the c_j are constants independent of t and q. Then we have:

$$\begin{aligned} ||\mathcal{E}_{\text{trap}}(t)||_{L_1((0,T))} &= \int_0^T \left|\frac{J_1(t)}{t} - \frac{2}{(q+1)\pi}\sum_{j=1}^{q}\sqrt{1-w_j^2}\cos w_j t\right| dt \\ &\leq \frac{1}{q}(c_1 T + c_2 T^2). \end{aligned} \tag{2.48}$$

Substituting this into (2.29) and using (2.46) we finally obtain, for some $\bar{c}(T)$ independent of q:

$$\int_0^T ||e(\cdot,t)||^2_{H_1[\Omega]} dt \leq \bar{c}\frac{T^4}{q}\int_0^T ||u(L,\cdot,t)||^2_{H_3[R^{n-1}]} dt. \tag{2.49}$$

We have shown, then, that the trapezoid rule produces a one-half order approximation to the true solution in the sense that the error decays at least as the square root of the reciprocal of the number of terms in the boundary condition.

A more accurate formula is obtained through the application of the Gaussian quadrature rule associated with the weight $\sqrt{1-w^2}$:

$$\frac{J_1(t)}{t} \approx \frac{1}{q+1}\sum_{i=1}^{q}\sin^2\frac{j\pi}{q+1}\cos w_j t, \quad w_j = \cos\frac{j\pi}{q+1}, \tag{2.50}$$

$$R(z) = \frac{1}{q+1} \sum_{j=1}^{q} \frac{(\sin^2 \frac{j\pi}{q+1}) z}{z^2 + w_j^2}. \tag{2.51}$$

This approximation is discussed in [15]. Remarkably, it is shown to be equivalent to the stable Padé approximants introduced by Engquist and Majda [7,8]. The stability of this approximation is well-known. To estimate the stability constant, we repeat the analysis given above for the trapezoid approximation. As in that case, we must estimate $(1 - |R(i)|)^{-1}$. We find:

$$(1 - |R(i)|)^{-1} = \left(1 - \frac{1}{q+1} \sum_{j=1}^{q} \frac{\sin^2 \frac{j\pi}{q+1}}{1 - w_j^2}\right)^{-1} = q + 1. \tag{2.52}$$

We also have:

$$\frac{1}{\sqrt{1 - w_1^2}} = \frac{1}{\sin \frac{\pi}{q+1}} \leq c(q+1). \tag{2.53}$$

Hence we conclude:

$$C \leq \bar{C}(T) q, \tag{2.54}$$

with $\bar{C}$ independent of q.

Using the error formula for the quadrature rule we find [15]:

$$||\mathcal{E}_{\text{gauss}}(t)||_{L_1(0,T)} \leq c \frac{(T/2)^{2q+1}}{(2q+1)!}. \tag{2.55}$$

From this, (2.29) and (2.54) we prove:

$$\int_0^T ||e(\cdot, t)||^2_{H_1[\Omega]} dt \leq \bar{c} \frac{q^2 (T/2)^{4q+2}}{((2q+1)!)^2} \int_0^T ||u(L, \cdot, t)||^2_{H_{2q+2}[R^{n-1}]} dt. \tag{2.56}$$

Here we see exponential convergence in q for smooth w, as with spectral approximations to the solutions of differential equations, so that the method might reasonably be called infinite order.

Clearly, these estimates are nonuniform in T, as must be the case for spatially local boundary conditions. (See (2.18).) In [14,15] we presented a number of methods for approximating the convolution kernel by decaying functions, including exponential interpolation, exponential least squares and approximations by Laguerre functions. None of these seemed entirely satisfactory: the interpolation could not be easily extended to high order, while the other approximations converged very slowly (if at all) in $L_1(0, \infty)$. (We note that the proposed conditions have not yet been tested.)

In order to use quadrature to construct long-time approximations we must derive an integral representation of $J_1(t)/t$ involving an integrand which decays in t. This may be accomplished by treating (2.4) as a complex integral and deforming the contour. For example, let:

$$z = w + i(1 - w^2)P(w), \quad P > 0, \quad w \in [-1, 1]. \tag{2.57}$$

Then:

$$\begin{aligned} \frac{J_1(t)}{t} &= \frac{1}{\pi}\Re\left(\int_C (1 - z^2)^{1/2} e^{izt} dz\right) \\ &= \frac{1}{2\pi}\int_{-1}^{1} \sqrt{1 - w^2} D(w, t) dw, \end{aligned} \tag{2.58}$$

$$D(w, t) = (f_1(w) \cos wt + f_2(w) \sin wt) e^{-(1-w^2)P(w)t}, \tag{2.59}$$

$$f_1(w) = g(w) + (2wP(w) - (1 - w^2)P'(w))h(w), \tag{2.60}$$

$$f_2(w) = (2wP(w) - (1 - w^2)P'(w))g(w) - h(w), \tag{2.61}$$

$$g(w) = G(w) + (1 - w^2)P^2(w)/G(w), \tag{2.62}$$

$$G(w) = \sqrt{1 + \sqrt{1 + (1 + w)^2 P^2(w)}}\sqrt{1 + \sqrt{1 + (1 - w)^2 P^2(w)}}, \tag{2.63}$$

$$h(w) = (1 - w)P(w)H(w) - (1 + w)P(w)/H(w), \tag{2.64}$$

$$H(w) = \frac{\sqrt{1 + \sqrt{1 + (1 + w)^2 P^2(w)}}}{\sqrt{1 + \sqrt{1 + (1 - w)^2 P^2(w)}}}. \tag{2.65}$$

Since the function $D(w, t)$ is smooth on the interval of integration, it is still reasonable to use the Gaussian quadrature scheme for the weight $\sqrt{1 - w^2}$. This yields:

$$\frac{J_1(t)}{t} \approx \frac{1}{2(q + 1)} \sum_{i=1}^{q} \sin^2 \frac{j\pi}{q + 1} D(w_j, t), \tag{2.66}$$

$$R(z) = \frac{1}{2(q + 1)} \sum_{j=1}^{q} r_j(z), \tag{2.67}$$

$$r_j(z) = \frac{\sin^2 \frac{j\pi}{q+1} (f_1(w_j)(z + (1 - w_j^2)P(w_j)) + f_2(w_j)w_j)}{(z + (1 - w_j^2)P(w_j))^2 + w_j^2}. \tag{2.68}$$

We have not yet tested or analyzed these schemes. Finite time error estimates should be obtainable using the error formula for the quadrature rule, but the hope is that time independent estimates will hold.

Problem B: Find a function or class of functions, $P(w)$, such that the resulting scheme leads to well-posed problems and kernels which converge to $\mathcal{K}(t)$ in $L_1[(0,\infty)]$.

3. Generalizations.

3.1. Anisotropic problems. The approximations discussed above are also applicable to problems with anisotropic wave propagation. As a first example, consider the convective wave equation with subsonic convection:

$$\left(\frac{\partial}{\partial t} + M\sum_l \omega_l \frac{\partial}{\partial x_l}\right)^2 u = \nabla^2 u, \quad x_1 \geq L. \tag{3.1}$$

We normalize so that $\sum_l \omega_l^2 = 1$, $0 \leq M < 1$. To formulate the exact boundary condition, we seek solutions of the form:

$$\hat{u} = Ae^{\lambda x_1}, \tag{3.2}$$

leading to the quadratic equation,

$$\left(s + iM\sum_{l>1} \omega_l k_l + M\omega_1\lambda\right)^2 = \lambda^2 - |k|^2. \tag{3.3}$$

The relevant solution, that is the one with negative real part for $\Re(s)$ sufficiently large, is given by:

$$\lambda = \tag{3.4}$$
$$-(1-M^2\omega_1^2)^{-1}\left((1-M\omega_1)\tilde{s} + \left((\tilde{s}^2 + (1-M^2\omega_1^2)|k|^2)^{1/2} - \tilde{s}\right)\right),$$

where

$$\tilde{s} = s + iM\sum_{l>1}\omega_l k_l. \tag{3.5}$$

To conveniently express approximations to this condition, we define the *tangential material derivative*, $D_{\tan}/Dt$, by

$$\frac{D_{\tan}}{Dt} = \frac{\partial}{\partial t} + U_{\tan}\cdot\nabla_{\tan}, \quad U_{\tan} = M\begin{pmatrix}\omega_2 & \omega_3 & \ldots & \omega_n\end{pmatrix}^T, \tag{3.6}$$

and let D/Dt denote the standard material derivative with respect to the full velocity $U = M(\omega_1 \ \ \ldots \ \ \omega_n)^T$. The exact condition is then given by:

$$(1+M\omega_1)\frac{\partial \bar{u}}{\partial x_1} + \frac{\partial \bar{u}}{\partial t} + i(U_{\tan}\cdot k)\bar{u} + (1+M\omega_1)|k|^2\mathcal{A}*\bar{u}, \tag{3.7}$$

$$A(k,t) = \frac{J_1(\sqrt{1-M^2\omega_1^2}|k|t}{\sqrt{1-M^2\omega_1^2}|k|t)} e^{-i(U_{\tan}\cdot k)t}. \tag{3.8}$$

Using the rational approximation $\sqrt{1-M^2\omega_1^2}|k|R(z)$, with $z = \tilde{s}/(\sqrt{1-M^2\omega_1^2}|k|)$, we obtain, in analogy with (2.11-2.12),

$$(1+M\omega_1)\frac{\partial \bar{v}}{\partial x_1} + \frac{\partial \bar{v}}{\partial t} + i(U_{\tan}\cdot k)\bar{v} + \sum_{j=1}^{p+1} \bar{h}_j = 0, \tag{3.9}$$

$$\left(\frac{\partial}{\partial t} + i(U_{\tan}\cdot k) - \sqrt{1-M^2\omega_1^2}|k|\rho_j\right)\bar{h}_j = \alpha_j|k|^2(1+M\omega_1)\bar{v}. \tag{3.10}$$

The approximate condition is local in space under the same conditions for locality in the isotropic case, that is (2.13). Then we have:

$$\frac{\partial v}{\partial x_1} + \frac{Dv}{Dt} + \sum_{j=1}^{q} \phi_j = 0, \tag{3.11}$$

$$\left(\frac{D_{\tan}^2}{Dt^2} - (1-M^2\omega_1^2)\beta_j^2\nabla_{\tan}^2\right)\phi_j = -(1+M\omega_1)\gamma_j\nabla_{\tan}^2\frac{D_{\tan}v}{Dt}. \tag{3.12}$$

We are confident that issues of consistency and convergence for these approximations could be handled as in the isotropic case, but we have not yet carried out the details. Below we see that the same operators appear in boundary conditions for the linearized compressible Euler and Navier-Stokes systems.

3.2. Applications to fluid dynamics. We now consider the compressible Euler equations linearized about a uniform flow in two space dimensions:

$$\frac{Du}{Dt} + \frac{\partial p}{\partial x} = 0, \tag{3.13}$$

$$\frac{Dv}{Dt} + \frac{\partial p}{\partial y} = 0, \tag{3.14}$$

$$\frac{Dp}{Dt} + \frac{\partial u}{\partial x} + \frac{\partial v}{\partial y} = 0, \tag{3.15}$$

where $D/Dt = \partial/\partial t + M\omega_x\partial/\partial x + M\omega_y\partial/\partial y$, $\omega_x^2 + \omega_y^2 = 1$, $0 < M < 1$. We suppose $\omega_x > 0$ and put artificial boundaries at $x = \pm L$; that is inflow at $-L$ and outflow at L.

To compute exact conditions we Laplace transform in t, Fourier transform in y, and rewrite the system in the form:

$$(3.16)\quad \frac{\partial}{\partial x}\begin{pmatrix}\hat{u}\\ \hat{v}\\ \hat{p}\end{pmatrix} = \begin{pmatrix} \frac{M\omega_x\tilde{s}}{1-M^2\omega_x^2} & -\frac{ik}{1-M^2\omega_x^2} & -\frac{\tilde{s}}{1-M^2\omega_x^2} \\ 0 & -\frac{\tilde{s}}{M\omega_x} & -\frac{ik}{M\omega_x} \\ -\frac{\tilde{s}}{1-M^2\omega_x^2} & \frac{ikM\omega_x}{1-M^2\omega_x^2} & \frac{M\omega_x\tilde{s}}{1-M^2\omega_x^2} \end{pmatrix}\begin{pmatrix}\hat{u}\\ \hat{v}\\ \hat{p}\end{pmatrix}.$$

Here, $\tilde{s} = s + ikM\omega_y$. Eigenvalues and left eigenvectors of the coefficient matrix are given by:

$$(3.17)\qquad \lambda_1 = \frac{\hat{A} + M\omega_x\tilde{s}}{1 - M^2\omega_x^2},\quad l_1^T = (\tilde{s} \quad -ikM\omega_x \quad -\hat{A}),$$

$$(3.18)\qquad \lambda_2 = -\frac{\hat{A} - M\omega_x\tilde{s}}{1 - M^2\omega_x^2},\quad l_2^T = (\tilde{s} \quad -ikM\omega_x \quad \hat{A}),$$

$$(3.19)\qquad \lambda_3 = -\frac{\tilde{s}}{M\omega_x},\quad l_3^T = (ikM\omega_x \quad \tilde{s} \quad ik),$$

where

$$(3.20)\qquad \hat{A} = (\tilde{s}^2 + (1 - M^2\omega_x^2)k^2)^{1/2}.$$

Noting that $\Re(\lambda_1) > 0$ and $\Re(\lambda_{2,3}) < 0$ for $\Re(s) > 0$ we have, setting $w = (u \quad v \quad p)^T$:

$$(3.21)\qquad l_1^T w = 0,\quad x = L;\quad l_2^T w = l_3^T w = 0,\quad x = -L.$$

The exact outflow boundary condition is then given by:

$$(3.22)\quad \left(\frac{\partial}{\partial t} + ikM\omega_y\right)(\bar{u} - \bar{p}) - ikM\omega_x\bar{v} - (1 - M^2\omega_x^2)k^2\mathcal{A} * \bar{p} = 0.$$

The exact inflow conditions are given by:

$$(3.23)\quad \left(\frac{\partial}{\partial t} + ikM\omega_y\right)(\bar{u} + \bar{p}) - ikM\omega_x\bar{v} + (1 - M^2\omega_x^2)k^2\mathcal{A} * \bar{p} = 0,$$

$$(3.24)\qquad \left(\frac{\partial}{\partial t} + ikM\omega_y\right)\bar{v} + ikM\omega_x\bar{u} + ik\bar{p} = 0.$$

Here,

$$(3.25)\qquad \mathcal{A}(k,t) = \frac{J_1(\sqrt{1 - M^2\omega_x^2}kt)}{\sqrt{1 - M^2\omega_x^2}kt}e^{-ikM\omega_y t}.$$

Using the rational approximation $\sqrt{1-M^2\omega_x^2}|k|R(z)$, for $z = \tilde{s}/(\sqrt{1-M^2\omega_x^2}|k|)$, equations (3.22- 3.23) become, in analogy with (3.9-3.10),

$$\left(\frac{\partial}{\partial t} + ikM\omega_y\right)(\bar{u}-\bar{p}) - ikM\omega_x\bar{v} - \sum_{j=1}^{p+1}\bar{h}_j = 0, \tag{3.26}$$

$$\left(\frac{\partial}{\partial t} + ikM\omega_y - \sqrt{1-M^2\omega_x^2}|k|\rho_j\right)\bar{h}_j = \alpha_j k^2(1-M^2\omega_x^2)\bar{p}, \tag{3.27}$$

at outflow and at inflow,

$$\left(\frac{\partial}{\partial t} + ikM\omega_y\right)(\bar{u}+\bar{p}) - ikM\omega_x\bar{v} + \sum_{j=1}^{p+1}\bar{g}_j = 0, \tag{3.28}$$

$$\left(\frac{\partial}{\partial t} + ikM\omega_y - \sqrt{1-M^2\omega_x^2}|k|\rho_j\right)\bar{g}_j = \alpha_j k^2(1-M^2\omega_x^2)\bar{p}. \tag{3.29}$$

The approximate conditions are local in space under the same conditions for locality in the scalar case, that is (2.13). In particular, if we use the Engquist-Majda-Padé-Gaussian approximation, (2.51), we have, at outflow:

$$\frac{D_{\tan}}{Dt}(u-p) - M\omega_x\frac{\partial v}{\partial y} - \sum_{j=1}^{q}\phi_j = 0, \tag{3.30}$$

$$\left(\frac{D_{\tan}}{Dt} + \sqrt{1-M^2\omega_x^2}\cos\frac{j\pi}{q+1}\frac{\partial}{\partial y}\right)\phi_j = -\frac{1}{q+1}\sin^2\frac{j\pi}{q+1}(1-M^2\omega_x^2)\frac{\partial^2 p}{\partial y^2}. \tag{3.31}$$

The inflow conditions become:

$$\frac{D_{\tan}}{Dt}(u+p) - M\omega_x\frac{\partial v}{\partial y} + \sum_{j=1}^{q}\psi_j = 0, \tag{3.32}$$

$$\left(\frac{D_{\tan}}{Dt} + \sqrt{1-M^2\omega_x^2}\cos\frac{j\pi}{q+1}\frac{\partial}{\partial y}\right)\psi_j = -\frac{1}{q+1}\sin^2\frac{j\pi}{q+1}(1-M^2\omega_x^2)\frac{\partial^2 p}{\partial y^2}. \tag{3.33}$$

$$\frac{D_{\tan}v}{Dt} + M\omega_x\frac{\partial u}{\partial y} + \frac{\partial p}{\partial y} = 0. \tag{3.34}$$

These conditions have been implemented by Goodrich [9] for channel flows, and shown to be very accurate even for moderate q. Note that (3.34) implies zero vorticity at inflow and, for the linearized problem, it is exact. Also, we have used an alternative representation of the approximate conditions with the property that the auxiliary functions are computed by solving first order equations.

A similar construction has been carried out for the linearized, isentropic, compressible Navier-Stokes equations in the low Mach number limit by the author and Lorenz [18]. Here, we require two boundary conditions at outflow and three at inflow. The conditions (3.22), (3.23) and (3.34) remain the same to leading order. The additional condition at outflow is

$$\frac{Dv}{Dt} + \frac{\partial p}{\partial y} = 0, \tag{3.35}$$

and at inflow is given by:

$$\frac{\partial u}{\partial t} + \frac{\partial p}{\partial x} = 0. \tag{3.36}$$

We note that here we generally expect that long time computations, as measured on the time scale of the sound waves, will be of interest. Therefore, nonlocal approximations may be efficient. As part of his doctoral dissertation, L. Xu is looking into such approximations and their applications in acoustics. Some of his results will appear in [19].

3.3. Corner conditions. The boundary conditions, as discussed so far, only apply to half-space or periodic problems. To generalize their applicability, one must understand how to treat the case of an artificial boundary intersecting another part of the boundary at a corner. Collino [4] has solved this problem for the important special case of the isotropic wave equation, two artificial boundaries intersecting at a right angle, and spatially local boundary conditions. (We note that our formulation of the exact conditions and spatially nonlocal approximations is not valid in this case.)

As a first attempt to generalize Collino's results, we have considered the convective wave equation in a rectangular domain with all boundaries artificial and the spatially local boundary conditions (3.11-3.12). Collino's construction does not directly apply, as it is based on special exact solutions of the wave equation in a corner. We have, instead, looked at power series expansions in the corner. These do lead to compatibility conditions which can be used to relate the auxiliary functions associated with distinct boundaries. However, it seems that the expansion must be carried out to order greater than $4q$ to produce the q required conditions. We have tried to do this symbolically, but so far have succeeded only for $q \leq 2$.

Problem C: Derive corner compatibility relations for local approximate boundary conditions for anisotropic systems.

We also note that the theory of exact conditions and nonlocal approximations is still undeveloped.

Problem D: Characterize the exact boundary condition for the wave equation and convective wave equation with a rectangle or rectangular parallelipiped as artificial boundary. Construct convergent temporally local approximate conditions.

Another case of practical importance should be analyzed is the intersection of an artificial boundary with a physical boundary, such as a solid wall. As a first example consider the convective wave equation in two space dimensions with convection in the x direction and walls at $y = 0$ and $y = H$:

$$\left(\frac{\partial}{\partial t} + M\frac{\partial}{\partial x}\right)^2 u = \nabla^2 u, \quad (x,y) \in (0,\infty) \times (0,H), \tag{3.37}$$

$$\alpha_0 \frac{\partial u}{\partial y} - \beta_0 u = 0, \quad y = 0; \quad \alpha_1 \frac{\partial u}{\partial y} + \beta_1 u = 0, \quad y = H. \tag{3.38}$$

We can expand the solution:

$$u = \sum_l \bar{u}_l Y_l(y), \quad Y_l(y) = A_l \cos k_l y + B_l \sin k_l y, \tag{3.39}$$

where the eigenvalues, k_l, are determined by the boundary conditions. Then, $\bar{u}_l$ satisfies (3.7), and for the approximate local boundary condition (3.11)-(3.12) we have:

$$\phi_j = \sum_l \bar{\phi}_{jl} Y_l(y). \tag{3.40}$$

Since the eigenfunctions, Y_l, satisfy (3.38) so would the ϕ_j, under the assumption that the expansion is sufficiently regular. That is:

$$\alpha_0 \frac{\partial \phi_j}{\partial y} - \beta_0 \phi_j = 0, \quad y = 0; \quad \alpha_1 \frac{\partial \phi_j}{\partial y} + \beta_1 \phi_j = 0, \quad y = H. \tag{3.41}$$

We should emphasize that this is generally not a true solution of the compatibility problem, but simply an approximation which we have used.

This approximation can be extended to the linearized, compressible Euler system, again with the physical assumption that the base flow is parallel to the walls, $y = 0, H$. The physical boundary condition at this characteristic boundary is:

$$v = 0, \tag{3.42}$$

while a second relation, implied by the y-momentum equation is:

$$\frac{\partial p}{\partial y} = 0. \tag{3.43}$$

Using these we see that the solution can be expanded in the form:

$$u = \sum_l \bar{u}_l \cos l\pi y/H, \quad v = \textstyle\sum_l \bar{v}_l \sin l\pi y/H,$$

$$p = \textstyle\sum_l \bar{p}_l \cos l\pi y/H. \tag{3.44}$$

To relate these expansions to expansions of the auxiliary variables, we note that:

$$\cos\frac{j\pi}{q+1} + \cos\frac{(q+1-j)\pi}{q+1} = 0. \tag{3.45}$$

Hence, equations (3.31) and (3.33) imply:

$$\phi_j + \phi_{q+1-j} = \sum_l \bar{\Phi}_l \cos l\pi y/H, \tag{3.46}$$

$$\psi_j + \psi_{q+1-j} = \sum_l \bar{\Psi}_l \cos l\pi y/H, \tag{3.47}$$

with boundary conditions:

$$\frac{\partial\phi_j}{\partial y} + \frac{\partial\phi_{q+1-j}}{\partial y} = 0, \quad \frac{\partial\psi_j}{\partial y} + \frac{\partial\psi_{q+1-j}}{\partial y} = 0, \quad y = 0, H. \tag{3.48}$$

These are the corner conditions used by Goodrich in [9].

3.4. Smooth boundaries. Given the theoretical and practical difficulties of high-order boundary conditions at artificial boundaries with corners, smooth artificial boundaries are an attractive alternative. Representations of the exact boundary condition for the wave equation at circular and spherical artificial boundaries are easily obtained. (See [17].)

Using polar coordinates in two dimensions, we place the artificial boundary at $r = R$ and expand the solution, u, in a Fourier series in θ:

$$u(r,\theta,t) = \sum_l \bar{u}_l e^{il\theta}. \tag{3.49}$$

The exact boundary condition is then given by:

$$\frac{\partial\bar{u}_l}{\partial r} + \frac{\partial\bar{u}_l}{\partial t} + \frac{1}{2R}\bar{u}_l + \frac{1}{R^2}A_l(t/R) * \bar{u}_l = 0, \tag{3.50}$$

$$\hat{A}_l(z) = -z\left(\frac{K_l'(z)}{K_l(z)} + 1 + \frac{1}{2z}\right). \tag{3.51}$$

(Here, $K_l(z)$ is the modified Bessel function [1].)

We have not, as yet, found a closed form expression for A_l. We note that conditions based on far-field expansions (e.g. [2,16]) correspond to large z expansions of $\hat{A}_l$. By the results of [17], long time accuracy is difficult to achieve using time-local conditions, as A_0 decays slowly as $t \to \infty$.

In three dimensions, on the other hand, the exact condition takes a much simpler form. Here we use spherical coordinates and take $r = R$ as our artificial boundary. The solution, u, is now expanded in spherical harmonics:

$$u(r,\theta,\phi,t) = \sum_l \bar{u}_l Y_l(\theta,\phi), \tag{3.52}$$

where

$$\nabla^2_{\text{sphere}} Y_l = -l(l+1)Y_l. \tag{3.53}$$

In analogy with the two-dimensional case, the exact condition is related to the inverse Laplace transform of the logarithmic derivative of spherical Bessel functions. In particular we have:

$$\frac{\partial \bar{u}_l}{\partial r} + \frac{\partial \bar{u}_l}{\partial t} + \frac{1}{R}\bar{u}_l + \frac{1}{R^2} S_l(t/R) * \bar{u}_l = 0, \tag{3.54}$$

$$\hat{S}_l(z) = -z\left(\frac{(z^{-1/2}K_{l+1/2}(z))'}{z^{-1/2}K_{l+1/2}(z)} + 1 + \frac{1}{z}\right) = \frac{P_l(z)}{Q_l(z)}, \tag{3.55}$$

where, for $l \neq 0$,

$$P_l(z) = \sum_{k=0}^{l-1} \frac{(2l-k)!}{k!(l-k-1)!}(2z)^k, \quad Q_l(z) = \sum_{k=0}^{l} \frac{(2l-k)!}{k!(l-k)!}(2z)^k, \tag{3.56}$$

and $S_0 = 0$.

We immediately observe that $S_l(z)$ is rational and, therefore, (3.54) may be localized in time. (The fact that the exact boundary condition can be localized in time for a finite number of spherical harmonics is also noted by Grote and Keller [10].) What has not been accomplished so far is to derive a convenient factorization or expansion of $P_l(z)/Q_l(z)$, to facilitate numerical implementation. However, even if this must be done numerically, the availability of simple, exact conditions makes the use of a spherical boundary very attractive.

We note that an important advantage of rectangular boundaries is the possibility of adjusting the aspect ratio of the domain. To do this with a smooth boundary, while maintaining a natural, separable coordinate system, one might need elliptical and spheroidal coordinates. This suggests our final problem.

Problem E: Characterize and approximate the exact boundary condition for the wave equation at elliptical and spheroidal boundaries.

Appendix.

A. Derivation of equation (2.4). Our goal is to compute a useful representation of the inverse Laplace transform of:

$$\hat{\mathcal{K}}(s) = \left(s^2+1\right)^{1/2} - s. \tag{A.1}$$

Although the inverse is available in standard tables, its importance to our work suggests the inclusion of a direct verification.

Following the ideas given in [5, Ch. 38], we will derive a simple differential equation satisfied by $\mathcal{K}(t)$. We first note that $\hat{\mathcal{K}}$ satisfies a first order differential equation with quadratic coefficients:

$$\left(s^2+1\right)\frac{d\hat{\mathcal{K}}}{ds} = s\hat{\mathcal{K}} - 1. \tag{A.2}$$

Recall that:

$$\frac{d\hat{f}}{ds} = \widehat{-tf(t)}, \quad s^p\hat{f} - \sum_{k=0}^{p-1} f^{(k)}(0^+)s^{p-k-1} = \widehat{f^{(p)}(t)}. \tag{A.3}$$

Therefore, (A.2) is equivalent to:

$$\widehat{\frac{d^2}{dt^2}(t\mathcal{K})} + \widehat{\frac{d\mathcal{K}}{dt}} + \widehat{t\mathcal{K}} = 1 - 2\mathcal{K}(0^+). \tag{A.4}$$

That is, $\mathcal{K}(t)$ satisfies the differential equation:

$$t\frac{d^2\mathcal{K}}{dt^2} + 3\frac{d\mathcal{K}}{dt} + t\mathcal{K} \equiv M\mathcal{K} = 0, \tag{A.5}$$

subject to the initial condition,

$$\mathcal{K}(0) = \frac{1}{2}, \tag{A.6}$$

in addition to a growth condition at infinity.

We now verify that:

$$\mathcal{K}(t) = \frac{1}{\pi}\int_{-1}^{1}\sqrt{1-w^2}\cos wt\,dw, \tag{A.7}$$

solves this problem. That the initial condition is satisfied may be checked directly. Applying the differential operator we obtain:

$$\begin{aligned} \pi M\mathcal{K} &= \int_{-1}^{1}(1-w^2)^{3/2}t\cos wt\,dw - 3\int_{-1}^{1}\sqrt{1-w^2}\,w\sin wt\,dw \\ &= \int_{-1}^{1}(1-w^2)^{3/2}t\cos wt\,dw + \int_{-1}^{1}\frac{d}{dw}(1-w^2)^{3/2}\sin wt\,dw \qquad \text{(A.8)} \\ &= 0. \end{aligned}$$

Finally, we recall the identity,

$$J_1(t) = \frac{t}{\pi}\int_{-1}^{1} \sqrt{1-w^2}\cos wt dw, \tag{A.9}$$

which may be found, e.g., in [1].

REFERENCES

[1] M. Abramowitz and I. Stegun, *Handbook of Mathematical Functions*, Dover (1972).

[2] A. Bayliss and E. Turkel, Radiation boundary conditions for wave-like equations, *Comm. Pure and Appl. Math.*, 33 (1980), 707-725.

[3] D. Braess, *Nonlinear Approximation Theory*, Springer-Verlag (1986).

[4] F. Collino, Conditions d'ordre élevé pour des modèles de propagation d'ondes dans des domaines rectangulaires, INRIA report 1790, (1993).

[5] G. Doetsch, *Introduction to the Theory and Application of the Laplace Transformation*, Springer-Verlag (1974).

[6] B. Engquist and L. Halpern, Far field boundary conditions for computation over long time, *Appl. Num. Math.*, 4 (1988), 21-45.

[7] B. Engquist and A. Majda, Absorbing boundary conditions for the numerical simulation of waves, *Math. Comp.*, 31 (1977), 629-651.

[8] B. Engquist and A. Majda, Radiation boundary conditions for acoustic and elastic wave calculations, *Comm. Pure and Appl. Math.*, 32 (1979), 313-357.

[9] J. Goodrich, An approach to the development of numerical algorithms for first order linear hyperbolic systems in multiple space dimensions: the constant coefficient case, preprint.

[10] M. Grote and J. Keller, Exact nonreflecting boundary conditions for the time-dependent wave equation, preprint.

[11] B. Gustafsson and H.-O. Kreiss, Boundary conditions for time-dependent problems with an artificial boundary, *J. Comp. Phys.*, 30 (1979), 333-351.

[12] T. Ha-Duong and P. Joly, On the stability analysis of boundary conditions for the wave equation by energy methods. Part I: The homogeneous case, *Math. Comp.*, 62 (1994), 539-564.

[13] T. Hagstrom, Asymptotic boundary conditions for dissipative waves: General theory, *Math. Comp.*, 56 (1991), 589-606.

[14] T. Hagstrom, Open boundary conditions for a parabolic system, *Math. Comput. Mod.*, 20 (1994), 55-68.

[15] T. Hagstrom, On the convergence of local approximations to pseudodifferential operators with applications, *Proc. of the 3rd Int. Conf. on Math. and Num. Aspects of Wave Prop. Phen.*, SIAM (1995), to appear.

[16] T. Hagstrom and S. Hariharan, Progressive wave expansions and open boundary problems, these Proceedings.

[17] T. Hagstrom, S. Hariharan and R. MacCamy, On the accurate long-time solution of the wave equation on exterior domains: Asymptotic expansions and corrected boundary conditions, *Math. Comp.*, 63 (1994), 507-539.

[18] T. Hagstrom and J. Lorenz, Boundary conditions and the simulation of low Mach number flows, *Proceedings of the First International Conference on Theoretical and Computational Acoustics*, D. Lee and M. Schultz, eds., 2 (1994), 657-668.

[19] T. Hagstrom, J. Lorenz and L. Xu, High-order radiation boundary conditions for low Mach number flows with applications, in preparation.

[20] L. Halpern and J. Rauch, Error analysis for absorbing boundary conditions, *Num. Math.*, 51 (1987), 459-467.

[21] H.-O. Kreiss and J. Lorenz, *Initial-Boundary Value Problems and the Navier-Stokes Equations*, Academic Press (1989).

[22] R. Sakamoto, *Hyperbolic Boundary Value Problems*, Cambridge University Press (1982).

[23] L. Trefethen and L. Halpern, Well-posedness of one-way wave equations and absorbing boundary conditions, *Math. Comp.*, 47 (1986), 421-435.

[24] L. Trefethen and L. Halpern, Wide-angle one-way wave equations, *J. Acoust. Soc. Am.*, 84 (1988), 1397-1404.

PROGRESSIVE WAVE EXPANSIONS AND OPEN BOUNDARY PROBLEMS

T. HAGSTROM* AND S.I. HARIHARAN†

Abstract. In this paper we construct progressive wave expansions and asymptotic boundary conditions for wave-like equations in exterior domains, including applications to electromagnetics, compressible flows and aero-acoustics. The development of the conditions will be discussed in two parts. The first part will include derivations of asymptotic conditions based on the well-known progressive wave expansions for the two-dimensional wave equations. A key feature in the derivations is that the resulting family of boundary conditions involve a single derivative in the direction normal to the open boundary. These conditions are easy to implement and an application in electromagnetics will be presented. The second part of the paper will discuss the theory for hyperbolic systems in two dimensions. Here, the focus will be to obtain the expansions in a general way and to use them to derive a class of boundary conditions that involve only time derivatives or time and tangential derivatives. Maxwell's equations and the compressible Euler equations are used as examples. Simulations with the linearized Euler equations are presented to validate the theory.

Key words. Progressive wave expansions, boundary conditions, Maxwell's Equations, Euler Equations, Numerical Simulations

AMS(MOS) subject classifications. 65M99, 35B40

1. Introduction. Exterior problems are commonly posed for wave-like equations, and their numerical solution leads to the problem of open boundary conditions. We discuss both isotropic and nonisotropic cases as they arise in electromagnetics and fluid dynamics. These equations include first order hyperbolic systems such as Maxwell's equations, the Euler equations of compressible flows, or the linearized Euler equations, as well as second order reduced forms as appropriate. Many work studies of this problem have appeared in the recent literature and we won't try to list them all. There are fundamentally two different, though related, approaches that have usually been taken. One is the use of high frequency asymptotics such as the geometrical optics approximation. The other is based on the far field structure of the solution. (For a third approach based on the direct approximation of the exact condition, see [5].) Progressive wave expansions were used as a tool to construct far field boundary conditions as early as the time of Sommerfeld. In the modern computational point of view, they were put in use for the first time by Kriegsmann and Morawetz [8]. Since then there have been many variations to this approach. For example, ex-

* Department of Mathematics and Statistics, The University of New Mexico, Albuquerque NM 87131. Supported in part by ICOMP, NASA Lewis Research Center, Cleveland, OH and NSF Grant No. DMS-9304406.

† Department of Mathematical Sciences, The University of Akron Akron, OH 44325-4002. Supported in part by a NASA cooperative agreement NCC3 - 283 and by ICOMP, NASA Lewis Research Center, Cleveland, OH.

tensions to anisotropic propagation were first attempted by Bayliss and Turkel [2], and a generalization to the case of anisotropic wave equations in two and three dimensions was proved by the authors [7]. Issues include the construction of the expansion for general systems, their use to construct stable boundary conditions of minimum order, and, finally, their practical implementation. Higher order conditions no matter which approach is used, are typically more complicated than the partial differential equation one starts with. In particular, they tend to have higher order derivatives in the direction of the propagation. To avoid this problem, often the partial differential equation itself is used. Such a procedure is not known in general for problems governed by first order hyperbolic systems. Here we provide a systematic way of dealing with this issue using progressive wave expansions. Our attention focuses on first order systems; namely, Maxwell's equations and the linearized Euler equations. To motivate the central ideas, we first consider the second order wave equation with the emphasis on progressive wave solutions.

2. Second order wave equation. As mentioned above, the goal here is to treat the problem of boundary conditions without having higher order normal derivatives. To illustrate the underlying procedure, let us consider the problem governed by the wave equation in two dimensions. We wish to construct the progressive wave solutions to this equation and exploit their structure to prescribe asymptotic boundary conditions. The equation written in cylindrical coordinates takes the form

$$u_{tt} = u_{rr} + \frac{1}{r}u_r + \frac{1}{r^2}u_{\theta\theta} \tag{2.1}$$

We look for solutions that are periodic in the angular direction as follows:

$$u(r,\theta,t) = \sum_{n=0}^{\infty} v_n(r,t)a_n(\theta) \tag{2.2}$$

where $a_n(\theta) = A_n \cos n\theta + B_n \sin n\theta$. Substituting (2.2) in (2.1), we obtain

$$v_{n,tt} = v_{n,rr} + \frac{1}{r}v_{n,r} - \frac{n^2}{r^2}v_n \tag{2.3}$$

Following Friedlander [4], we construct solutions of (2.3) in the form

$$v_n(r,t) = \sum_{j=0}^{\infty} \frac{f_j^n(t-r)}{r^{j+\frac{1}{2}}} \tag{2.4}$$

Substitution of (2.4) in (2.3) results in the following recurrence relations

$$f_{j+1}^{n'}(t-r) = -\frac{(j+\frac{1}{2})^2 - n^2}{2(j+1)} f_j^n(t-r) \tag{2.5}$$

The goal is to examine the effect of the recurrence relations on constructions of asymptotic boundary conditions. First, we observe that substitution of (2.4) in (2.2) yields the following formal representation of the solution:

$$u(r,\theta,t) = \sum_{n=0}^{\infty} a_n(\theta) \sum_{j=0}^{\infty} \frac{f_j^n(t-r)}{r^{j+\frac{1}{2}}} \tag{2.6}$$

Manipulations of this series, particularly increasing the order of the decay rate for boundary conditions, have been proposed by many authors (e.g. Bayliss and Turkel [1]). In fact, a different form of (2.6) has been used for these constructions, which will not be discussed here. We define a "basic boundary operator" from (2.6) as follows:

$$B := \frac{\partial}{\partial t} + \frac{\partial}{\partial r} + \frac{1}{2r} \tag{2.7}$$

It is immediately verified from (2.6) that

$$Bu = -\sum_{n=0}^{\infty} a_n(\theta) \sum_{j=1}^{\infty} \frac{j f_j^n(t-r)}{r^{j+\frac{3}{2}}} \tag{2.8}$$

Direct approximation of (2.8) is the radiation condition $Bu = 0$, a popular condition in the literature noted by many researchers (e.g. Bayliss and Turkel [1], Engquist and Majda [3]). Asymptotic accuracy of such a condition is $0(r^{-5/2})$, which is evident from (2.8). Higher order conditions in general require higher order normal derivatives or derivatives in the dominant direction of propagation. This may not be a desirable feature numerically, particularly for nonlinear generalizations. Here we obtain higher order conditions that involve Bu, u, $u_{\theta\theta}$, and their time derivatives on the artificial boundary. We begin with the construction of higher order conditions by differentiating (2.8). This yields:

$$\frac{\partial}{\partial t} Bu = -\sum_{n=0}^{\infty} a_n(\theta) \sum_{j=1}^{\infty} \frac{j f_j^{n\prime}(t-r)}{r^{j+\frac{3}{2}}} \tag{2.9}$$

Noting that the inner summation may be written in the form

$$\sum_{j=0}^{\infty} \frac{(j+1) f_{j+1}^{n\prime}(t-r)}{r^{j+\frac{5}{2}}}$$

and using the recurrence relation (2.5) yields:

$$\frac{\partial}{\partial t} Bu = \sum_{n=0}^{\infty} a_n(\theta) \sum_{j=0}^{\infty} \frac{(j+\frac{1}{2})^2 - n^2}{2r^{j+\frac{5}{2}}} f_j^n$$

A simple manipulation of the right hand side yields:

$$\frac{\partial}{\partial t}Bu - \frac{1}{8r^2}u - \frac{1}{2r^2}\frac{\partial^2 u}{\partial\theta^2} = \frac{1}{2}\sum_{n=0}^{\infty} a_n(\theta)\sum_{j=1}^{\infty}\frac{j(j+1)}{r^{j+\frac{5}{2}}}f_j^n \tag{2.10}$$

The highlight here is the observation

$$u_{\theta\theta} = -\sum_{n=0}^{\infty} n^2 a_n(\theta)\sum_{j=0}^{\infty}\frac{f_j^n(t-r)}{r^{j+\frac{1}{2}}} \tag{2.11}$$

Note that the asymptotic accuracy of the candidate boundary condition (2.10) is increased further to $0(r^{-\frac{7}{2}})$. Let

$$B_1 u = \frac{\partial}{\partial t}Bu - \frac{u}{8r^2} - \frac{u_{\theta\theta}}{r^2} \tag{2.12}$$

Then (2.10) becomes

$$B_1 u = \frac{1}{2}\sum_{n=0}^{\infty} a_n(\theta)\sum_{j=0}^{\infty}\frac{(j+1)(j+2)}{r^{j+\frac{7}{2}}}f_{j+1}^n \tag{2.13}$$

This form again suggests the use of the recurrence relations (2.5) by differentiating the equation with respect to time. Doing so, we obtain

$$\frac{\partial}{\partial t}B_1 u = -\frac{1}{4}\sum_{n=0}^{\infty} a_n(\theta)\sum_{j=0}^{\infty}\frac{(j+2)[(j+\frac{1}{2})^2 - n^2]}{r^{j+\frac{7}{2}}}f_j^n \tag{2.14}$$

which is equivalent to

$$\frac{\partial}{\partial t}B_1 u + \frac{1}{8r^3}u + \frac{1}{2r^3}u_{\theta\theta} = -\frac{1}{4}\sum_{n=0}^{\infty} a_n(\theta)\sum_{j=1}^{\infty}\frac{j[(j+\frac{3}{2})^2 - n^2]}{r^{j+\frac{7}{2}}}f_j^n . \tag{2.15}$$

We note that $-n^2$ translates into the second tangential derivative. Defining

$$B_2 u = \frac{\partial}{\partial t}B_1 u + \frac{1}{8r^3}u + \frac{1}{2r^3}u_{\theta\theta} \tag{2.16}$$

it is clear that equation (2.15) yields a one asymptotic order higher boundary condition (to $O(r^{-\frac{9}{2}})$). Moreover, noting

$$B_2 u = -\frac{1}{4}\sum_{n=0}^{\infty} a_n(\theta)\sum_{j=0}^{\infty}\frac{(j+1)[(j+\frac{5}{2})^2 - n^2]}{r^{j+\frac{9}{2}}}f_{j+1}^n \tag{2.17}$$

and applying the time derivative again, the process becomes clear and it yields

$$B_3 u := \frac{\partial}{\partial t}B_2 u - \frac{25}{128r^4}u - \frac{13}{16r^4}u_{\theta\theta} - \frac{1}{8r^4}u_{\theta\theta\theta\theta} = O(\frac{1}{r^{\frac{11}{2}}}). \tag{2.18}$$

Remark 1: As far as numerical implementation of these conditions are concerned, one may consider a sequence of operations to update u at the current time:

$$u_t + u_r + \frac{1}{2r}u = z \tag{2.19}$$

$$z_t - \frac{1}{8r^2}u + \frac{1}{r^2}u_{\theta\theta} = v \tag{2.20}$$

$$v_t + \frac{1}{8r^3}u + \frac{1}{2r^3}u_{\theta\theta} = w \tag{2.21}$$

$$w_t - \frac{25}{128r^4}u - \frac{13}{16r^4}u_{\theta\theta} - \frac{1}{8r^4}u_{\theta\theta\theta\theta} = 0 \tag{2.22}$$

The above sequence of equations (which provides a boundary condition asymptotically accurate to $O(r^{-\frac{11}{2}})$) as a system of first order equations to march in time. (At the time this article was written one of the students of the second author has implemented such a procedure and obtained the indicated asymptotic improvement. The details will appear elsewhere).

Remark 2: The procedure above coincides with the high frequency approximations of the exact condition in the radially symmetric case. In the Laplace transform domain, the exact operator has the form (see [6])

$$B_e u = -\frac{1}{r}\frac{srK_0'(sr)}{K_0(sr)}. \tag{2.23}$$

Where $K_0(z)$ is the modified Bessel function of order 0. Moreover we find as $sr \to \infty$

$$B_e u = \frac{1}{r}(sr + \frac{1}{2} - \frac{1}{8sr} + \frac{1}{8(sr)^2} + O((sr)^{-3})) \tag{2.24}$$

The Laplace transform of the derived operators coincides with the large sr approximations of the exact boundary operator B_e. We can, then, interpret the expansions both as a long-range and as a high-frequency approximation.

We also note that the Fourier transform of the operator $B_1 u$ coincides with the second order operator proposed by Kriegsmann et al. [9] in conjunction with on surface radiation conditions. As an example we consider the computation of the surface current calculation in electromagnetic scattering. Let Γ be the boundary of a perfect conductor. Then the magnitude of the total current is given by the formula (see [9])

$$J = |\frac{i}{k}\frac{\partial}{\partial n}(u_s + u_{inc})|_\Gamma \tag{2.25}$$

where u_s is the scattered field, and u_{inc} is the known incident field. For perfect conductors $u_s = -u_{inc}$ on the boundary of the scatterer Γ. The principle of the on surface boundary procedure consists of bringing the far field boundary exactly on the interface of the scatterer. The advantage is rather clear. Since the total current is a functional of the normal derivative of the scattered field and the radiation boundary operators on the surface directly express the normal derivatives in terms of the incident field. We note that in the formula for the surface current k is the wave number which arose from the Fourier transform of the wave equation. We list the Fourier transform of the operators derived in our theory. They are:
Condition 1.

$$-iku + u_r + \frac{1}{2r}u = 0, \tag{2.26}$$

Condition 2.

$$-ik(-iku + u_r + \frac{1}{2r}u) = \frac{1}{8r^2} + \frac{1}{2r^2}u_{\theta\theta}, \tag{2.27}$$

Condition 3.

$$(-ik)^2(-iku + u_r + \frac{1}{2r}u) = -ik(\frac{1}{8r^2}u + \frac{1}{2r^2}u_{\theta\theta}) - \frac{1}{8r^3}u - \frac{1}{2r^3}u_{\theta\theta}. \tag{2.28}$$

The first two operators are used in [9], and the third one is, so far as we know, new. A plane wave incident upon a unit cylinder is considered for the calculation of J and results are shown in Figures 2.1 ($k = 5$) and 2.2 ($k = 2$) respectively. The incident field has the specific form $u_{inc} = e^{ikr\cos\theta}$.

Remark 3.: For anisotropic equations, such as convective wave equation, an analogous procedure may be derived. The use of the resulting conditions are more pertinent to systems of equations such as the linearized Euler equation. This is discussed in section 4.

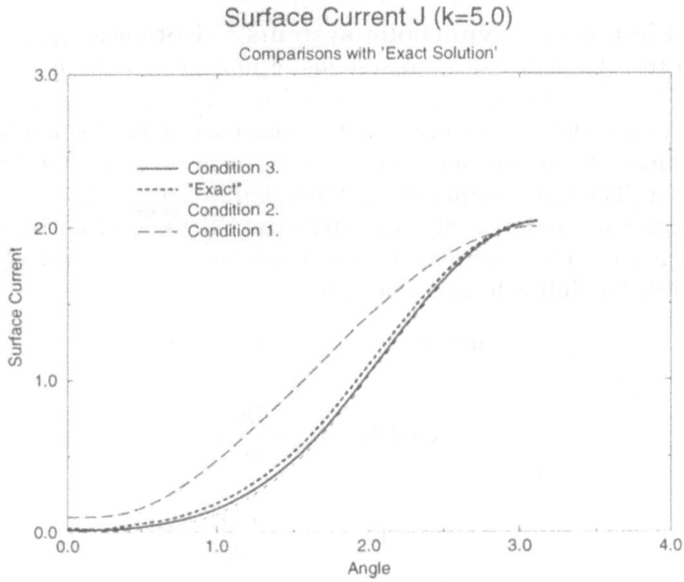

FIG. 2.1. *Comparison of results with exact solution,* $k = 5$

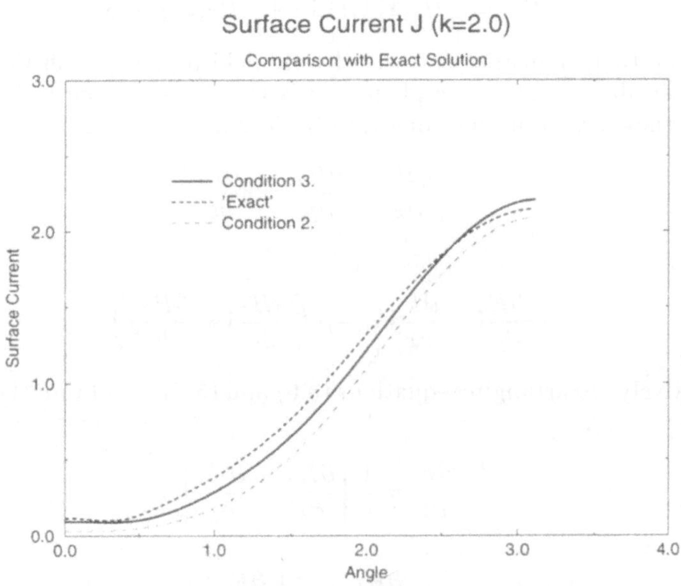

FIG. 2.2. *Comparison of results with exact solution,* $k = 2$

3. First order hyperbolic systems – isotropic case. Here our focus is to extend the ideas to systems of first order equations. The progressive wave expansions may be carried out directly in the time domain as we did for the second order wave equations or in the Laplace transform domain. In this section we present the construction using the Laplace transform. The direct approach is illustrated in Section 4.

Maxwell's equations offer an interesting example of an isotropic system. Here we confine our attention to Transverse Magnetic (TM) fields for simplicity. The full field equations are:

$$div\ \epsilon\mathbf{E} = div\ \mu\mathbf{H} = 0, \tag{3.1}$$

$$curl\ \mathbf{H} = \epsilon\frac{\partial \mathbf{E}}{\partial t}, \tag{3.2}$$

$$curl\ \mathbf{E} = -\mu\frac{\partial \mathbf{H}}{\partial t}. \tag{3.3}$$

We shall consider TM fields as follows:

$$\mathbf{E} = E(x,y,t)\,\mathbf{k} \tag{3.4}$$

$$\mathbf{H} = H_1(x,y,t)\,\mathbf{i} + H_2(x,y,t)\,\mathbf{j}. \tag{3.5}$$

Equation (3.4) indicates that the electric field propagates in the direction perpendicular to the $x-y$ plane and is transverse to the magnetic field. Under these assumptions, equations (3.2) and (3.3) become

$$\frac{\partial H_2}{\partial x} - \frac{\partial H_1}{\partial y} = \epsilon\frac{\partial E}{\partial t} \tag{3.6}$$

and

$$\frac{\partial E}{\partial y}\mathbf{i} - \frac{\partial E}{\partial x}\mathbf{j} = -\mu\left(\frac{\partial H_1}{\partial t}\mathbf{i} + \frac{\partial H_2}{\partial t}\mathbf{j}\right) \tag{3.7}$$

respectively. Rearranging equations (3.6) and (3.7), we obtain the following system:

$$\frac{\partial E}{\partial t} = \frac{1}{\epsilon}\left[\frac{\partial H_2}{\partial x} - \frac{\partial H_1}{\partial y}\right]$$

$$\frac{\partial H_1}{\partial t} = -\frac{1}{\mu}\frac{\partial E}{\partial y}$$

$$\frac{\partial H_2}{\partial t} = \frac{1}{\mu}\frac{\partial E}{\partial x}$$

This can be put in the conventional form:

$$\mathbf{u}_t = \mathbf{A}\ \mathbf{u}_x + \mathbf{B}\ \mathbf{u}_y \tag{3.8}$$

where

$$\mathbf{A} = \begin{pmatrix} 0 & 0 & \frac{1}{\epsilon} \\ 0 & 0 & 0 \\ \frac{1}{\mu} & 0 & 0 \end{pmatrix},$$

$$\mathbf{B} = \begin{pmatrix} 0 & -\frac{1}{\epsilon} & 0 \\ -\frac{1}{\mu} & 0 & 0 \\ 0 & 0 & 0 \end{pmatrix}$$

and where $\mathbf{u} = (E, H_1, H_2)^T$. Converting to polar coordinates we obtain:

$$\frac{\partial}{\partial t}\begin{pmatrix} E \\ H_1 \\ H_2 \end{pmatrix} = \begin{pmatrix} 0 & -\frac{1}{\epsilon}\sin\theta & \frac{1}{\epsilon}\cos\theta \\ -\frac{1}{\mu}\sin\theta & 0 & 0 \\ \frac{1}{\mu}\cos\theta & 0 & 0 \end{pmatrix} \frac{\partial}{\partial r}\begin{pmatrix} E \\ H_1 \\ H_2 \end{pmatrix}$$

$$+\frac{1}{r}\begin{pmatrix} 0 & -\frac{1}{\epsilon}\cos\theta & -\frac{1}{\epsilon}\sin\theta \\ -\frac{1}{\mu}\cos\theta & 0 & 0 \\ -\frac{1}{\mu}\sin\theta & 0 & 0 \end{pmatrix} \frac{\partial}{\partial\theta}\begin{pmatrix} E \\ H_1 \\ H_2 \end{pmatrix} \tag{3.9}$$

We take the Laplace transform of (3.9). With the change of variable $\tilde{r} = rs$ we have

$$\begin{pmatrix} \hat{E} \\ \hat{H}_1 \\ \hat{H}_2 \end{pmatrix} = \left(R\frac{\partial}{\partial\tilde{r}} + \frac{1}{\tilde{r}}\Theta\frac{\partial}{\partial\theta} \right) \begin{pmatrix} \hat{E} \\ \hat{H}_1 \\ \hat{H}_2 \end{pmatrix} \tag{3.10}$$

where

$$R = \begin{pmatrix} 0 & -\frac{1}{\epsilon}\sin\theta & \frac{1}{\epsilon}\cos\theta \\ -\frac{1}{\mu}\sin\theta & 0 & 0 \\ \frac{1}{\mu}\cos\theta & 0 & 0 \end{pmatrix}$$

and

$$\Theta = \begin{pmatrix} 0 & -\frac{1}{\epsilon}\cos\theta & -\frac{1}{\epsilon}\sin\theta \\ -\frac{1}{\mu}\cos\theta & 0 & 0 \\ -\frac{1}{\mu}\sin\theta & 0 & 0 \end{pmatrix}$$

We seek an expansion of solutions of (3.10) in the form

$$\begin{pmatrix} \hat{E} \\ \hat{H}_1 \\ \hat{H}_2 \end{pmatrix} = \frac{e^{-\tilde{r}g(\theta)}}{\tilde{r}^{\alpha}} \left(\mathbf{a} + \frac{1}{\tilde{r}}\mathbf{b} + \cdots \right) \tag{3.11}$$

We note that this form is similar to Friedlander's form that applies to the second order wave equation in the Laplace transform domain. Also, we have introduced a decay rate constant α which turns out in two dimensions to equal $\frac{1}{2}$, as expected. Substituting (3.11) into (3.10) yields to leading order:

$$A = I + gR + g'\Theta \tag{3.12}$$

and

$$A\mathbf{a} = \mathbf{0} \tag{3.13}$$

For this requirement to be true, clearly it must follow that:

$$0 = det(A) = det \begin{pmatrix} 1 & -\frac{1}{\epsilon}(g\sin\theta + g'\cos\theta) & \frac{1}{\epsilon}(g\cos\theta - g'\sin\theta) \\ -\frac{1}{\mu}(g\sin\theta + g'\cos\theta) & 1 & 0 \\ \frac{1}{\mu}(g\cos\theta - g'\sin\theta) & 0 & 1 \end{pmatrix}$$

$$= (1 - \frac{1}{\epsilon\mu}((g\sin\theta + g'\cos\theta)^2 + (g\cos\theta - g'\sin\theta)^2))$$

$$= 1 - \frac{1}{\epsilon\mu}(g^2 + (g')^2) \tag{3.14}$$

The roots of equation (3.14) are:

$$g = \pm\sqrt{\epsilon\mu}, \sqrt{\epsilon\mu}\cos(\theta + \phi),$$

ϕ arbitrary. For waves propagating to infinity in all directions we choose $g = \sqrt{\epsilon\mu}$ as the allowoable root. With this value of g, the matrix A becomes

$$A = \begin{pmatrix} 1 & -\sqrt{\frac{\mu}{\epsilon}}\sin\theta & \sqrt{\frac{\mu}{\epsilon}}\cos\theta \\ -\sqrt{\frac{\epsilon}{\mu}}\sin\theta & 1 & 0 \\ \sqrt{\frac{\epsilon}{\mu}}\cos\theta & 0 & 1 \end{pmatrix}$$

whose right nullvector is

$$\mathbf{a} = \begin{pmatrix} 1 \\ \sqrt{\frac{\epsilon}{\mu}}\sin\theta \\ -\sqrt{\frac{\epsilon}{\mu}}\cos\theta \end{pmatrix} a_1(\theta)$$

with left nullvector

$$\mathbf{l}^T = (1, \sqrt{\frac{\mu}{\epsilon}}\sin\theta, -\sqrt{\frac{\mu}{\epsilon}}\cos\theta)$$

The next order terms in the asymptotic expansion yield the following relation:

$$A\mathbf{b} = -\alpha R\mathbf{a} + \Theta\frac{\partial \mathbf{a}}{\partial \theta}, \tag{3.15}$$

and α is determined by:

$$\alpha = \frac{\mathbf{1}^T \Theta \frac{\partial \mathbf{a}}{\partial \theta}}{\mathbf{1}^T R\mathbf{a}}, \tag{3.16}$$

i.e., by requiring $(\mathbf{1}, A\mathbf{b}) = 0$.

Noting the following calculations:

$$R\mathbf{a} = a_1(\theta)\begin{pmatrix} -\frac{1}{\sqrt{\epsilon\mu}} \\ -\frac{1}{\mu}\sin\theta \\ \frac{1}{\mu}\cos\theta \end{pmatrix},$$

$$\frac{\partial \mathbf{a}}{\partial \theta} = \frac{\partial a_1}{\partial \theta}\begin{pmatrix} 1 \\ \sqrt{\frac{\epsilon}{\mu}}\sin\theta \\ -\sqrt{\frac{\epsilon}{\mu}}\cos\theta \end{pmatrix} + a_1\begin{pmatrix} 0 \\ \sqrt{\frac{\epsilon}{\mu}}\cos\theta \\ \sqrt{\frac{\epsilon}{\mu}}\sin\theta \end{pmatrix},$$

$$\Theta\frac{\partial \mathbf{a}}{\partial \theta} = \frac{\partial a_1}{\partial \theta}\begin{pmatrix} 0 \\ -\frac{1}{\mu}\cos\theta \\ -\frac{1}{\mu}\sin\theta \end{pmatrix} + a_1\begin{pmatrix} -\frac{1}{\sqrt{\epsilon\mu}} \\ 0 \\ 0 \end{pmatrix},$$

$$\mathbf{1}^T R\mathbf{a} = a_1(\theta)(\frac{-2}{\sqrt{\epsilon\mu}}), \quad \mathbf{1}^T\Theta\frac{\partial \mathbf{a}}{\partial \theta} = a_1(\theta)(\frac{-1}{\sqrt{\epsilon\mu}}),$$

it follows that

$$\alpha = \frac{1}{2}.$$

Choose

$$\mathbf{b} = \begin{pmatrix} 0 \\ b_2 \\ b_3 \end{pmatrix},$$

and use the last two equations to obtain

$$b_2 = (\frac{1}{2\mu}\sin\theta)a_1 - \frac{1}{\mu}\cos\theta\frac{\partial a_1}{\partial \theta},$$

$$b_3 = (-\frac{1}{2\mu}\cos\theta)a_1 - \frac{1}{\mu}\sin\theta\frac{\partial a_1}{\partial\theta}.$$

Substituting **a** and **b** in (3.11) we obtain

$$\hat{E} = \frac{e^{-\tilde{r}\sqrt{\mu\epsilon}}}{\tilde{r}^{\frac{1}{2}}}a_1, \quad \frac{\partial\hat{E}}{\partial\theta} = \frac{e^{-\tilde{r}\sqrt{\mu\epsilon}}}{\tilde{r}^{\frac{1}{2}}}\frac{\partial a_1}{\partial\theta},$$

$$\hat{H}_1 = \sqrt{\frac{\epsilon}{\mu}}\sin\theta\hat{E} + \frac{1}{\mu rs}(\frac{\sin\theta}{2}\hat{E} - \cos\theta\frac{\partial\hat{E}}{\partial\theta}),$$

$$\hat{H}_2 = -\sqrt{\frac{\epsilon}{\mu}}\cos\theta\hat{E} + \frac{1}{\mu rs}(\frac{-\cos\theta}{2}\hat{E} - \sin\theta\frac{\partial\hat{E}}{\partial\theta}).$$

Multiplying through by s and taking the inverse transform finally yields:

$$\frac{\partial H_1}{\partial t} = (\sqrt{\frac{\epsilon}{\mu}}\sin\theta\frac{\partial}{\partial t} + \frac{\sin\theta}{2\mu r} - \frac{\cos\theta}{\mu r}\frac{\partial}{\partial\theta})E, \tag{3.17}$$

$$\frac{\partial H_2}{\partial t} = -(\sqrt{\frac{\epsilon}{\mu}}\cos\theta\frac{\partial}{\partial t} + \frac{\cos\theta}{2\mu r} + \frac{\sin\theta}{\mu r}\frac{\partial}{\partial\theta})E. \tag{3.18}$$

As only one boundary condition is required, we convert these into a single condition. Multiplying (3.17) $\sin\theta$ and subtracting (3.18) multiplied by $\cos\theta$, we obtain our final form:

$$\frac{\partial H_1}{\partial t}\sin\theta - \frac{\partial H_2}{\partial t}\cos\theta - \sqrt{\frac{\epsilon}{\mu}}\frac{\partial E}{\partial t} = \frac{1}{2\mu r}E \tag{3.19}$$

This construction is easily extended to higher order, though we have not devised a unified approach to the implementation of the higher order conditons.

4. The linearized Euler equations – an anisotropic example. The construction of asymptotic boundary conditions for the linearized and nonlinear compressible Euler equations is also of interest, particularly for applications in aeroacoustics. In this section, we construct the expansions in the time domain directly. Again, the system takes the form:

$$\mathbf{u}_t = \mathbf{A}\ \mathbf{u}_x + \mathbf{B}\ \mathbf{u}_y \tag{4.1}$$

where **A** and **B** are constant matrices. In cylindrical coordinates we have:

$$\mathbf{u}_t = \mathbf{R}\ \mathbf{u}_r + \frac{1}{r}\mathbf{T}\ \mathbf{u}_\theta, \tag{4.2}$$

where

$$\mathbf{R} = \mathbf{A}\cos\theta + \mathbf{B}\sin\theta, \tag{4.3}$$

$$\mathbf{T} = -\mathbf{A}\sin\theta + \mathbf{B}\cos\theta. \tag{4.4}$$

A far field asymptotic expansion may be constructed in the following form for the solution vector $\mathbf{u}$:

$$\mathbf{u} = \sum_{n=0}^{\infty} \frac{f_n(t - rg(\theta))}{r^{n+\frac{1}{2}}}\mathbf{a}_n, \tag{4.5}$$

where the scalar function $g(\theta)$ and the vectors $\mathbf{a}_n(\theta)$ are to be determined. The function f_0 is analogous to the radiation function discussed in [4]. The other functions are recursively determined by substitution of the expansion into equation (4.2). The $O(\frac{1}{r})$ terms yield:

$$\mathbf{C}\mathbf{a}_0 = \mathbf{0}, \tag{4.6}$$

where $\mathbf{C} = \mathbf{I} + g(\theta)\mathbf{R} + g'(\theta)\mathbf{T}$. For $\mathbf{a}_0$ to be nontrivial one must have $det(\mathbf{C}) = 0$, yielding an 'eikonal function' $g(\theta)$. The next order correction yields

$$f_0\mathbf{C}\mathbf{a}_1 = -f_1'\frac{1}{2}\mathbf{R}\mathbf{a}_0 + \mathbf{T}\frac{\partial \mathbf{a}_0}{\partial \theta}. \tag{4.7}$$

This imposes a necessary restriction that $f_0 = f_1'$. In general, it follows $f_{n-1} = f_n', n \geq 1$. At this point, we turn to the isentropic, linearized, compressible, Euler equations to illustrate the actual calculations involved in solving these algebraic problems. For a uniform base flow in the x direction they are:

$$(\frac{\partial}{\partial t} + M\frac{\partial}{\partial x})p + \frac{\partial u}{\partial x} + \frac{\partial v}{\partial y} = 0, \tag{4.8}$$

$$(\frac{\partial}{\partial t} + M\frac{\partial}{\partial x})u + \frac{\partial p}{\partial x} = 0, \tag{4.9}$$

$$(\frac{\partial}{\partial t} + M\frac{\partial}{\partial x})v + \frac{\partial p}{\partial y} = 0. \tag{4.10}$$

Conversion of this system to cylindrical coordinates (4.8)-(4.10) takes the form (4.2) where

$$\mathbf{u} = \begin{pmatrix} p \\ u \\ v \end{pmatrix},$$

$$\mathbf{R} = \begin{pmatrix} M\cos\theta & \cos\theta & \sin\theta \\ \cos\theta & M\cos\theta & 0 \\ \sin\theta & 0 & M\cos\theta \end{pmatrix},$$

$$\mathbf{T} = \begin{pmatrix} -M\sin\theta & -\sin\theta & \cos\theta \\ -\sin\theta & -M\sin\theta & 0 \\ \cos\theta & 0 & -M\sin\theta \end{pmatrix}.$$

Calculation of $g(\theta)$ for these equations yield:

$$g(\theta) = \frac{1}{\sqrt{1 - M^2\sin^2\theta} + M\cos\theta}, \tag{4.11}$$

and the matrix $\mathbf{C}$ has the form

$$\mathbf{C} = \begin{bmatrix} 1 - MQ & -Q & -R \\ -Q & 1 - MQ & 0 \\ -R & 0 & 1 - MQ \end{bmatrix},$$

where

$$Q = g\cos\theta - g'\sin\theta,$$

$$R = g\sin\theta + g'\cos\theta.$$

Solutions of (4.6) are given by

$$\mathbf{a}_0 = h_0(\theta)\begin{pmatrix} 1 \\ \frac{Q}{1-MQ} \\ \frac{R}{1-MQ} \end{pmatrix} = h_0(\theta)\begin{pmatrix} 1 \\ r_2 \\ r_3 \end{pmatrix},$$

and the solutions of (4.7) are given by:

$$\mathbf{a}_1 = h_1(\theta)\begin{pmatrix} 1 \\ r_2 \\ r_3 \end{pmatrix} + h_0(\theta)\begin{pmatrix} 0 \\ b_2 \\ b_3 \end{pmatrix} + h_0'(\theta)\begin{pmatrix} 0 \\ c_2 \\ c_3 \end{pmatrix}. \tag{4.12}$$

Here $h_0(\theta)$ and $h_1(\theta)$ are arbitrary functions of θ and the coefficients b_i and c_i are given by

$$b_2 = \frac{(\cos\theta + Mr_2\cos\theta)/2 + Mr_2'\sin\theta}{1 - MQ},$$

$$b_3 = \frac{(\sin\theta + Mr_3\cos\theta)/2 + Mr_3'\sin\theta}{1 - MQ},$$

$$c_2 = \frac{\sin\theta + M r_2 \sin\theta}{1 - MQ},$$

$$c_3 = \frac{-\cos\theta + M r_3 \sin\theta}{1 - MQ}.$$

Collection of these results in the asymptotic expansions yields (to $O(r^{-5/2})$):

$$p = \frac{h_0 f_0}{r^{1/2}} + \frac{h_1 f_1}{r^{3/2}}, \tag{4.13}$$

$$u = r_2 p + \frac{h_0 f_1}{r^{3/2}} b_2 + \frac{h_0' f_1}{r^{3/2}} c_2, \tag{4.14}$$

$$v = r_3 p + \frac{h_0 f_1}{r^{3/2}} b_3 + \frac{h_0' f_1}{r^{3/2}} c_3. \tag{4.15}$$

Differentiating u and v with respect to t and using the result $f_0 = f_1'$ to $O(r^{-5/2})$, we have

$$u_t = r_2 p_t + \frac{p}{r} b_2 + \frac{h_0' f_0}{r^{3/2}} c_2, \tag{4.16}$$

$$v_t = r_3 p_t + \frac{p}{r} r_3 + \frac{h_0' f_0}{r^{3/2}} c_3. \tag{4.17}$$

Finally, noting the term involving h_0' can be eliminated from the last two equations, we have

$$\alpha\, p_t + \beta\, u_t + \gamma\, v_t = \frac{p}{r}\delta, \tag{4.18}$$

where

$$\alpha = r_2 c_3 - r_3 c_2,$$

$$\beta = -c_3,$$

$$\gamma = c_2,$$

$$\delta = b_3 c_2 - b_2 c_3.$$

The higher order conditions are obtained in a similar manner. In fact, one can show that the next order condition is of the form

$$(\alpha\, p + \beta\, u + \gamma\, v)_{tt} = \frac{p_t}{r}\delta + \frac{p}{r^2}\epsilon, \tag{4.19}$$

which is accurate to $O(r^{-7/2})$. Note that these conditions do not involve any spatial derivatives. As such they are ideal for rectangular domains where typically one has to pay special attention to the corners, particularly when high order numerical schemes are used. These conditions correspond to the primary acoustic boundary condition. In addition, one must impose at inflow boundaries, a vorticity condition. For nonisentropic flows, in addition to the vorticity, entropy must also be specified at inflow. At inflow the y momentum equation and zero vorticity condition yield an exact relation:

$$v_t + (p + Mu)_y = 0. \tag{4.20}$$

5. A model problem. In this section we begin with the linearized Euler equations with mean velocity convection. The scaled form of these equations are identical to the one that we used to derive the conditions, except they contain forcing terms that characterize a driving source. They are:

$$(\frac{\partial}{\partial t} + M\frac{\partial}{\partial x})p + \frac{\partial u}{\partial x} + \frac{\partial v}{\partial y} = 0, \tag{5.1}$$

$$(\frac{\partial}{\partial t} + M\frac{\partial}{\partial x})u + \frac{\partial p}{\partial x} = g_1(x, y, t), \tag{5.2}$$

$$(\frac{\partial}{\partial t} + M\frac{\partial}{\partial x})v + \frac{\partial p}{\partial y} = g_2(x, y, t), \tag{5.3}$$

where $g1$ and $g2$ model a Gaussian momentum source, which both oscillates sinusoidally and decays algebraically in time. Typical examples include, a quadrupole sound distribution. Here g_1 and g_2 are the gradient of a potential ϕ. Such a function is given by

$$\phi = A(t)e^{-\alpha R^2} cos(2\theta)$$

where $\tan\theta = \frac{y-y_0}{x-x_0}$, $R = \sqrt{(x-x_0)^2 + (y-y_0)^2}$, (x_0, y_0) is the location of the source, α is a positive constant, and $A(t)$ is the amplitude and a function of t. (In the numerical experiments $A = \sin 2\pi t/(1+t^2)$.)

In a sample computation which was computed for a time length of 100 periods (22415 time steps), the solution obtained with the second order conditions was compared with the exact solution, a solution obtained by setting incoming characteristic variables to zero, and one obtained using the first order condition. The exact solution was computed in a large domain in which, within the time of computations, the waves could not reflect off the artificial boundaries and return to the small domain. The maximum error in pressure calculations observed for the characteristic conditions was 10.3%, for the first order asymptotic condition it was 3.3%, and for the second order condition the error was 1.3%; indicating the expected improvement. In Figures 5.1 and 5.2, the exact solution for the pressure is given after 5 periods and 10 periods of time respectively. Subsequent pairs of figures (5.3-5.4, 5.5-5.6 and 5.7-5.8) indicate the solution at these times for the characteristic boundary condition, the first order asymptotic condition, and the second order asymptotic conditions respectively. At 10 periods the errors are visible in the first two cases and their orders of the magnitude indeed are as indicated above. Clearly the higher order condition improved the results.

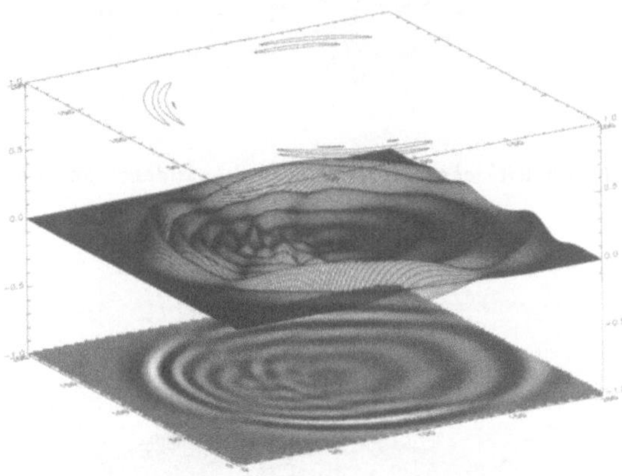

FIG. 5.1. *Exact Solution at 5 periods of time*

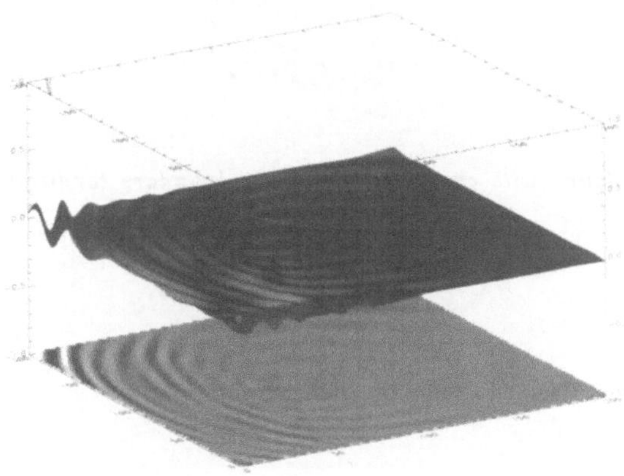

FIG. 5.2. *Exact Solution at 10 periods of time*

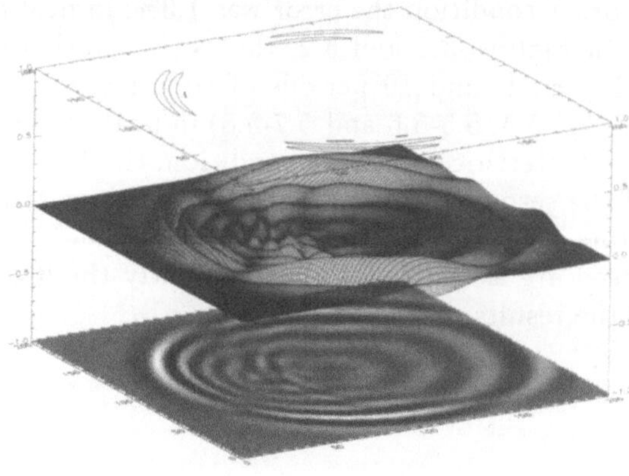

FIG. 5.3. *Solution with characteristics based boundary condition t = 5 periods*

FIG. 5.4. *Solution with characteristics based boundary condition t = 10 periods*

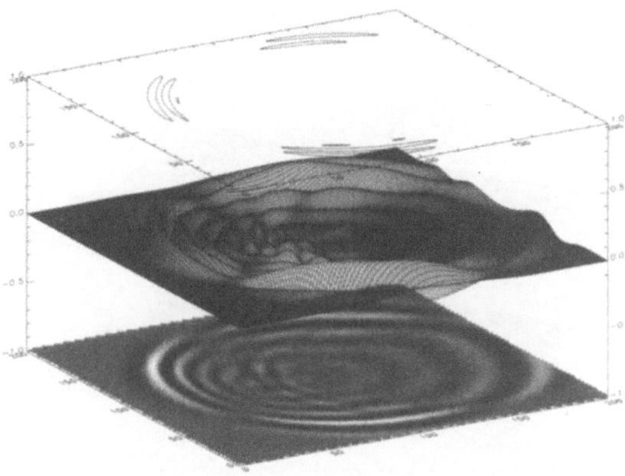

FIG. 5.5. *Solution with first asymptotic boundary condition $t = 5$ periods*

FIG. 5.6. *Solution with first asymptotic boundary condition $t = 10$ periods*

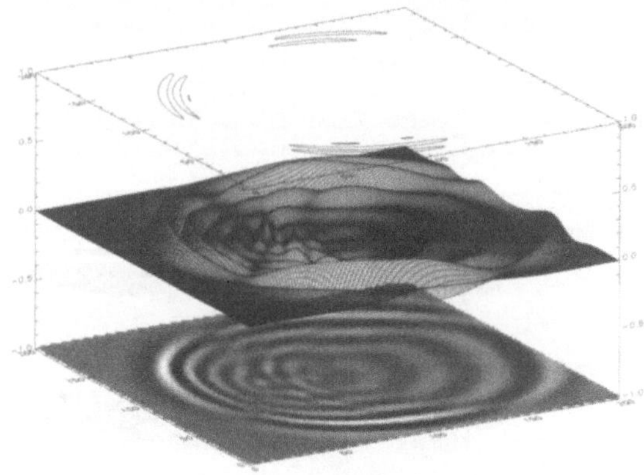

FIG. 5.7. *Solution with second asymptotic boundary condition $t = 5$ periods*

FIG. 5.8. *Solution with second asymptotic boundary condition $t = 10$ periods*

REFERENCES

[1] A. Bayliss and E. Turkel, "Radiation boundary conditions for wave like equations", *Commun. Pure Appl. Math.*, vol. 23, (1980) pp. 707-725.

[2] A. Bayliss and E. Turkel, "Far field boundary conditions for compressible flows", *J. Comp. Phys.*, 48, (1982), 182-199.

[3] B. Engquist and A. Majda, "Absorbing boundary conditions for the numerical simulation of waves", *Math. Comp.*, 31, (1977), pp. 629-651.

[4] F.G. Friedlander, "On the radiation field of pulse solutions of the wave equation", *Proc. Roy. Soc.* A, 279, (1964), pp. 386-394.

[5] T. Hagstrom, "On high-order radiation boundary conditions", this volume.

[6] T. Hagstrom, S. I. Hariharan, R. C. MacCamy, "On the accurate long-time solution of the wave equation in exterior domains: Asymptotic expansions and corrected boundary conditions", *Math. Comp.* , Vol. 63, No. 208, (1994) pp. 507-539.

[7] S. I. Hariharan and T. Hagstrom, "Far field expansion for anisotropic wave equations," *Proceedings of the Second IMACS Symposium on Computational Acoustics*, North Holland, Eds. D. Lee, A. Cakmak and R. Vichnevetsky, (1990), pp. 283-294.

[8] G. A. Kriegsmann and C. S. Morawetz, "Solving the Helmholtz equation for exterior problems with variable index of refraction:I", *SIAM J. Stat. Comput.*, Vol. 1, No. 3, (1980), pp. 371-385.

[9] G.A. Kriegsmann, A. Taflove, and K. R. Umashakar, "A new formulation of electromagnetic wave scattering using an on-surface radiation boundary condition approach", *IEEE Trans. Antennas Propagat.*, vol. AP-35, no. 2, pp. 153-161, 1987.

FORMULATION OF SPHERICAL NEAR-FIELD SCANNING IN THE TIME DOMAIN

THORKILD B. HANSEN*

Abstract. Spherical near-field scanning techniques are formulated for acoustic and electromagnetic fields in the time domain so that a single set of time-domain near-field measurements yields the far field in the time domain or over a wide range of frequencies. Probe-corrected as well as non-probe-corrected formulations are presented. For bandlimited time-domain fields, sampling theorems and computation schemes are derived that give the field outside the scan sphere in terms of sampled near-field data.

1. Introduction and summary of results. Near-field scanning techniques have been used extensively over the past 20 years to accurately determine fields from scatterers and antennas [1], [2]. With near-field scanning techniques the near field of the scatterer or antenna is first measured on a scan surface. Then these near-field measurements are used along with a set of near-field formulas to compute the desired field of the scatterer or antenna. One of the most important applications of near-field techniques has been the determination of far fields of highly specialized antennas from measurements of their near fields taken in radio anechoic chambers.

All near-field techniques rely on near-field formulas that express the field outside the scan surface in terms of the field on the scan surface or in terms of the output of a probe on the scan surface. The near-field formulas without probe correction require that the field on the scan surface is known, whereas the probe-corrected formulas require knowledge only of the output of a probe on the scan surface. The probe-corrected formulas express the field outside the scan surface in terms of the receiving characteristic of the probe and the probe output on the scan surface.

Three different scan surfaces in the near field have been used extensively: the planar surface [3], [4], the cylindrical surface [5], [6], and the spherical surface [7]-[12]. Spherical near-field scanning, which is the topic of the present paper, was first formulated in 1970 in the thesis of Jensen [7]. Jensen derived a probe-corrected spherical near-field formula that expresses the field of an antenna (called the test antenna), located inside the scan sphere, in terms of the receiving characteristic of the probe and the probe output on the scan sphere. Wacker [8] then showed how a special symmetric probe could be used to deconvolve Jensen's formulas to give a practical set of spherical near-field formulas. This set of near-field formulas was then used by Larsen [9] in 1977 to perform the first complete spherical near-field measurements with probe correction. Based on this work, Hansen

* Rome Laboratory ERCT, Hanscom AirForce Base, 31 Grenier Street, Hanscom, MA 01731-3010. This work was supported by the Danish Technical Research Council, Copenhagen, Denmark and by the Air Force Office of Scientific Research, Bolling AFB, DC.

et al. wrote a comprehensive book [10], that deals with the theoretical as well as the practical aspects of spherical near-field scanning. In this book, a key element in the derivation of the probe-corrected spherical near-field formulas is the use of the complicated translation and rotation formulas for spherical vector-wave functions. These translation and rotation formulas are used to determine the receiving coefficients of the probe for spherical waves emanating from the test antenna, given the receiving coefficients of the probe for spherical waves in a coordinate system centered on the probe.

Subsequently, Yaghjian and Wittmann [11], [12] found an ingenious method for deriving new probe-corrected spherical near-field formulas. Their method begins by expressing the output of an arbitrary probe as a linear spatial differential operator acting on the incident field at a reference point of the probe. Using this operator expression for the output of the probe, Yaghjian and Wittmann were able to derive probe-corrected spherical near-field formulas that do not involve rotation or translation theorems for the spherical wave functions. Moreover, their formulation casts the probe-corrected spherical near-field formulas in a simpler form which is similar to that of the corresponding formulas without probe correction.

All of this work with spherical near-field scanning has been limited to the frequency domain, so that radiation or scattering is determined for one frequency at a time. With the advent of broad-band radar it has become increasingly important to be able to accurately compute the fields of antennas and scatterers excited by short pulses. Because conventional frequency-domain methods are not always efficient in dealing with broad-band signals, it is appropriate to extend the spherical near-field techniques to the time domain.

This paper addresses the problem of formulating spherical near-field scanning for antennas in the time domain, so that a single set of time-domain near-field measurements on the scan sphere yields the time-domain field everywhere outside the scan sphere (in particular the far-field). In previous work [13]-[17], we successfully extended planar near-field scanning techniques to the time domain and extensive use will be made of these previous results here. In the present paper we present brief derivations of the spherical near-field formulas for both acoustic and electromagnetic fields. First formulas without probe correction are derived and then Yaghjian and Wittmann's operator formulation is used to determine the corresponding probe-corrected formulas.

In Section 2 we present the formulas without probe correction for both acoustic and electromagnetic fields. We begin in Section 2.1 with the acoustic field and derive a formula, which is based on a spherical eigenfunction expansion, that expresses the time-domain acoustic field outside the scan sphere in terms of its values on the scan sphere. Similarly, in Section 2.2 a formula is derived that expresses the time-domain electromagnetic field outside the scan sphere in terms of the time-domain tangential electric field on the scan sphere. This electromagnetic formula is based on a spherical eigenfunction expansion involving the transverse spherical vector-wave functions.

The probe-corrected time-domain spherical near-field formulas are derived in Section 3 for both acoustic and electromagnetic fields. The derivations are performed using the eigenfunction expansions of Section 2 and the operator formulation of Yaghjian and Wittmann [11], [12].

We start in Section 3.1 with a presentation of the operator formula for acoustic fields that expresses the frequency-domain output of an arbitrary probe as a linear spatial differential operator acting on the incident field at some reference point of the probe. Then, we derive the time-domain probe-corrected formulas for the acoustic field with the assumption that the probe is axis symmetric, that is, its receiving characteristic is independent of the azimuthal angle ϕ. The final time-domain probe-corrected formulas are in the same form as the corresponding formulas without probe correction. A special case of an axis-symmetric time-derivative probe is also considered.

In Section 3.2 we begin by presenting the operator formula for electromagnetic fields that expresses the output of an arbitrary probe as a linear spatial differential operator acting on the incident electric and magnetic fields at some reference point of the probe. Then, we derive the time-domain probe-corrected formulas for the electromagnetic field under the assumption that the probe is rotationally symmetric and that its receiving characteristic has only $\cos\phi$ and $\sin\phi$ dependence. (Such a probe is called a rotationally symmetric $\mu = \pm 1$ probe in [10, p.151].) As for the acoustic field, the final electromagnetic time-domain probe-corrected formulas are in the same form as the corresponding formulas without probe correction.

The spherical near-field formulas derived in Sections 2 and 3 involve infinite summations and require that the field be measured at all points on the scan sphere. In Section 4 it is shown that when the time-domain fields are bandlimited, the infinite summations can be truncated and the fields need only be measured on the scan sphere at a finite number of space-time points. Formulas are derived in Sections 4.1 and 4.2 that express the acoustic and electromagnetic fields, respectively, outside the scan sphere in terms of sampled near-field data on the scan sphere. For the formulas without probe correction the sampled near-field data consist of space-time sampled values of the field on the scan sphere. Similarly, for the probe-corrected formulas the sampled near-field data consist of space-time sampled values of the probe output on the scan sphere.

In Section 4.3 a numerical example illustrates the use of the acoustic eigenfunction formulas without probe correction. In this example the field is generated by an acoustic point source with Gaussian time dependence. We use results from the previous work on time-domain planar near-field scanning [13], [16] to determine the effective bandlimit of this source and then use the sampling theorems and formulas of Section 4.1 to compute the acoustic field outside the scan sphere. The results obtained this way are then compared with the exact expression for the point-source field. It is found that when the space-time sample spacing is in accordance with the sampling theorem, the spherical near-field formulas indeed give the correct field. Also, because the spherical scan surface is not truncated (as was the planar scan surface [13, ch. 4], [15]) no erroneous signal occurs in the computed field. Thus, in this respect time-domain spherical near-field scanning is advantageous over time-domain planar near-field scanning. However, the time-domain spherical near-field formulas are much more complicated than the corresponding planar formulas and thus more work is required to implement time-domain spherical near-field scanning.

2. Formulas without probe correction. We shall now derive the time-domain spherical near-field formulas without probe correction that give the acoustic and electromagnetic fields outside the scan sphere in terms of their values on the scan sphere. The starting point of the derivation for the acoustic field is the frequency-domain spherical eigenfunction expansion that expresses the field outside a source region in terms of spherical harmonics and spherical Hankel functions. This frequency-domain expansion is then Fourier transformed to get the corresponding time-domain expansion. Similarly, for the electromagnetic field the time-domain formula is obtained by Fourier transforming the frequency-domain spherical vector-wave function expansion.

The spherical scanning geometry shown in Figure 1 will be considered. The arbitrary source region of finite extent is located inside the minimum sphere of radius r_s and the values of the fields are assumed to be known on the scan sphere of radius a. In the acoustic case, the part of space not occupied by the sources is a linear, lossless fluid so that the time-domain acoustic pressure satisfies the scalar homogeneous wave equation outside the source region. In the electromagnetic case, the part of space not occupied by the sources is lossless free space with permeability μ and permittivity ϵ. In addition to the rectangular coordinates (x, y, z), the usual spherical coordinates (r, θ, ϕ) defined by $x = r\cos\phi\sin\theta$, $y = r\sin\phi\sin\theta$, and $z = r\cos\theta$ will also be used.

Throughout, $e^{-i\omega t}$ time dependence is suppressed in all the frequency-domain equations and all frequency-domain fields are labeled with ω. Furthermore, all time-domain fields outside and on the scan sphere (that is, in the region $r \geq a$) are assumed to be zero for $t < t_0$.

2.1. Acoustic fields. The time-domain acoustic field is denoted by $\Phi(\bar{r}, t)$ and satisfies the scalar homogeneous wave equation outside the source region, that is

$$\nabla^2\Phi(\bar{r}, t) - \frac{1}{c^2}\frac{\partial^2}{\partial t^2}\Phi(\bar{r}, t) = 0, \quad r > r_s \tag{2.1}$$

where c is the wave speed. We can construct the time-domain field $\Phi(\bar{r}, t)$ from the corresponding frequency-domain field $\Phi_\omega(\bar{r})$ (and vice versa) by the Fourier transform identities

$$(2.2)\ \Phi(\bar{r},t) = \int_{-\infty}^{+\infty} \Phi_\omega(\bar{r})e^{-i\omega t}d\omega, \quad \Phi_\omega(\bar{r}) = \frac{1}{2\pi}\int_{-\infty}^{+\infty} \Phi(\bar{r},t)e^{i\omega t}dt$$

and the frequency-domain acoustic field Φ_ω satisfies the scalar Helmholtz equation and the radiation conditions [18, p.57].

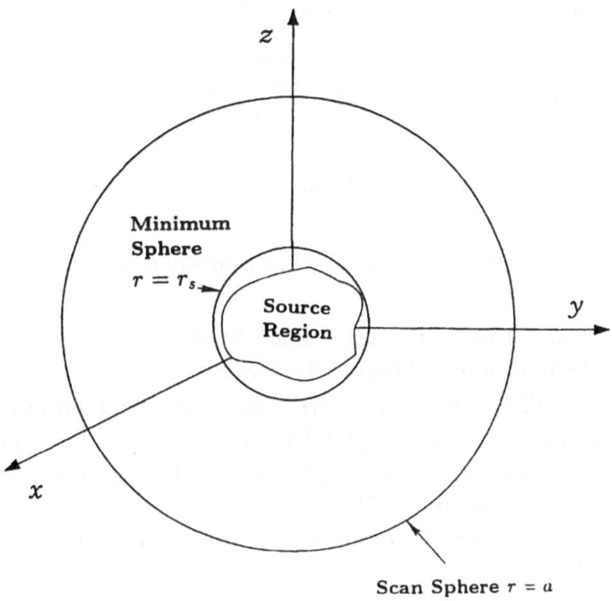

FIG. 1. *Spherical scanning geometry.*

The near-field formulas of this section are based on the well-known spherical-harmonics expansion of the frequency-domain acoustic field [18, sec.8.11]. For the subsequent time-domain derivation we write this expansion as [1]

$$(2.3) \qquad \Phi_\omega(\bar{r}) = \sum_{n=0}^{\infty} \frac{h_n^{(1)}(kr)}{h_n^{(1)}(ka)} \sum_{m=-n}^{n} C_{nm}^{\omega} Y_{nm}(\theta,\phi), \quad r \geq a$$

[1] Although the expansion (2.3) was derived for $\omega > 0$ it is easily shown to be valid for all real ω. To show this begin by noting that the spectrum Φ_ω of any real acoustic field $\Phi(t)$ must satisfy $\Phi_{-\omega} = \Phi_\omega^*$. Since $h_n^{(1)}(-\omega r/c)/h_n^{(1)}(-\omega a/c) = \left[h_n^{(1)}(\omega r/c)/h_n^{(1)}(\omega a/c)\right]^*$ and $C_{nm}^{-\omega}Y_{nm} = \left(C_{n,-m}^{\omega}Y_{n,-m}\right)^*$ we see that the spectrum defined by (2.3) indeed satisfies $\Phi_{-\omega} = \Phi_\omega^*$ which in turn shows that (2.3) is valid for all real omega.

where $k = \omega/c$ is the wave number and the expansion coefficients are determined by the acoustic field on the scan sphere by

$$C^{\omega}_{nm} = \int_0^{2\pi}\int_0^{\pi} \Phi_\omega(a\hat{r})Y^*_{nm}(\theta,\phi)\sin\theta d\theta d\phi. \tag{2.4}$$

Here $h_n^{(1)}(z)$ is the spherical Hankel function of the first kind defined in [19, p.439], $*$ denotes complex conjugation, and $Y_{nm}(\theta,\phi)$ are the normalized orthogonal spherical harmonics given by [20, p.99]

$$Y_{nm}(\theta,\phi) = \sqrt{\frac{2n+1}{4\pi}\frac{(n-m)!}{(n+m)!}}P_n^m(\cos\theta)e^{im\phi} \tag{2.5}$$

satisfying the orthogonality relations

$$\int_0^{2\pi}\int_0^{\pi} Y_{nm}(\theta,\phi)Y^*_{qs}(\theta,\phi)\sin\theta d\theta d\phi = \delta_{nq}\delta_{ms} \tag{2.6}$$

where δ_{nq} is the Kronecker delta defined by $\delta_{nq} = 1$ for $q = n$ and $\delta_{nq} = 0$ for $q \neq n$. The functions $P_n^m(\cos\theta)$ are the associated Legendre functions as defined in [20]. The factor $(h_n^{(1)}(ka))^{-1}$ has been included in the expansion (2.3) to cancel the singularity of $h_n^{(1)}(kr)$ at $k = 0$ and to make the time-domain expansion coefficients $C_{nm}(t)$ simple. As can be seen directly from (2.4), these time-domain coefficients are given by the integral over the scan sphere of the time-domain acoustic field multiplied by the spherical harmonics, that is

$$C_{nm}(t) = \int_0^{2\pi}\int_0^{\pi} \Phi(a\hat{r},t)Y^*_{nm}(\theta,\phi)\sin\theta d\theta d\phi. \tag{2.7}$$

From the frequency-domain expansion (2.3) we see that the time-domain acoustic field can be determined as a time convolution of the expansion coefficient (2.7) and the time function

$$F_n(r,t) = \int_{-\infty}^{+\infty} \frac{h_n^{(1)}(kr)}{h_n^{(1)}(ka)}e^{-i\omega t}d\omega, \tag{2.8}$$

that is,

$$\Phi(\bar{r},t) = \frac{1}{2\pi}\sum_{n=0}^{\infty}\sum_{m=-n}^{n}\int_{-\infty}^{+\infty} F_n(r,t')C_{nm}(t-t')dt'Y_{nm}(\theta,\phi), \quad r \geq a \tag{2.9}$$

and we will now derive an explicit expression for the time function F_n so that the convolution integral in (2.9) can be evaluated. This expression will be derived by the use of complex function theory and since the spherical Hankel function is given by [19, p.439]

$$h_n^{(1)}(z) = i^{-n-1}\frac{e^{iz}}{z}\sum_{q=0}^{n}(n+\frac{1}{2},q)(-2iz)^{-q} \tag{2.10}$$

the only possible singularities of the integrand in (2.8) in the complex ω plane are at the zeros of the spherical Hankel function. According to [21, sec.9], [19, pp.441, 373] there are n such zeros and we denote them by ω_{ns} and have

$$h_n^{(1)}(a\omega_{ns}/c) = 0, \quad s = 1, 2, 3, ..., n. \tag{2.11}$$

All these zeros have negative imaginary part and are located inside the circle of radius cn/a, that is $|a\omega_{ns}/c| < n$, [21, sec.9], [19, pp.441, 373] as shown in Figure 2. The zeros of the spherical Hankel functions $h_n^{(1)}(z)$ for $n = 1, 2, ..., 12$ are listed in [22, app.B]. Moreover, in [22, app.B] and [23] an asymptotic formula is presented that determines these zeros for large n.

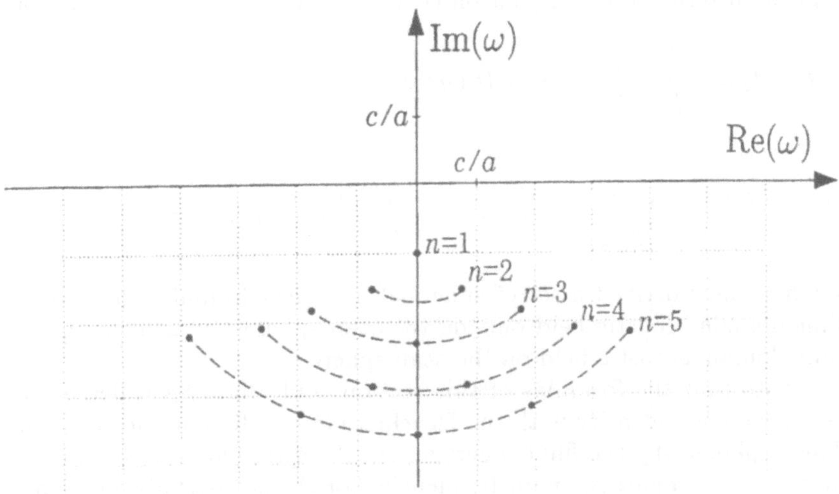

FIG. 2. *The zeros ω_{ns} determined by $h_n^{(1)}(a\omega_{ns}/c) = 0$.*

The integral in (2.8) is easily calculated by making use of the residue theorem and the asymptotic form of the spherical Hankel function to get [22, sec.2.1], [24]

$$F_n(r,t) = \frac{2\pi a}{r}\delta(t-r/c+a/c) - u(t-r/c+a/c)\frac{2\pi i c}{a}\sum_{s=1}^{n}\frac{h_n^{(1)}(\omega_{ns}r/c)}{h_n^{(1)\prime}(\omega_{ns}a/c)}e^{-i\omega_{ns}t} \tag{2.12}$$

where $h_n^{(1)'}(z)$ is derivative of the spherical Hankel function with respect to its argument and $u(t)$ is the unit step function given by $u(t) = 0$ for $t \leq 0$ and $u(t) = 1$ for $t > 0$.

Substituting the expression (2.12) for the function $F_n(r,t)$ into the time-convolution integral in (2.9) produces the near-field formula for the time-domain acoustic field

$$\Phi(\bar{r},t) = \sum_{n=0}^{\infty} \sum_{m=-n}^{n} Y_{nm}(\theta,\phi) \Bigg\{ \frac{a}{r} C_{nm}(t - r/c + a/c)$$

$$- \frac{ic}{a} \sum_{s=1}^{n} \frac{h_n^{(1)}(\omega_{ns} r/c)}{h_n^{(1)'}(\omega_{ns} a/c)} \int_{t_0}^{t-r/c+a/c} C_{nm}(t') e^{-i\omega_{ns}(t-t')} dt' \Bigg\}, \quad r \geq a \tag{2.13}$$

where we have used the fact that all fields in the region $r \geq a$ are zero for $t < t_0$. Using the property that the zeros ω_{ns} lie symmetrically with respect to the imaginary axis one may show that the right side of (2.13) indeed is real.

From the formula (2.13) it is seen that the time-domain far-field pattern $\mathcal{F}(\theta,\phi,t)$, which is defined such that $\Phi(\bar{r},t) \sim \mathcal{F}(\theta,\phi,t-r/c)/r$ for $r \to \infty$, is given in terms of the expansion coefficients $C_{nm}(t)$ by the expression

$$\mathcal{F}(\theta,\phi,t) = a \sum_{n=0}^{\infty} \sum_{m=-n}^{n} \Bigg\{ C_{nm}(t+a/c)$$

$$- \frac{c^2(-i)^n}{a^2} \sum_{s=1}^{n} \frac{e^{-i\omega_{ns} t}}{\omega_{ns} h_n^{(1)'}(\omega_{ns} a/c)} \int_{t_0}^{t+a/c} C_{nm}(t') e^{i\omega_{ns} t'} dt' \Bigg\} Y_{nm}(\theta,\phi). \tag{2.14}$$

We have now derived a set of spherical near-field formulas that gives the time-domain acoustic field outside the scan sphere $r = a$ in terms of the time-domain acoustic field on the scan sphere.

Note that the formulas of this section could have been derived from Green's second identity with the Dirichlet Green's function for the sphere. This explains why the final spherical near-field formulas (2.13) and (2.14) involve the complex resonant frequencies for the acoustically soft sphere.

2.2. Electromagnetic fields. In this section we derive the time-domain spherical near-field formulas without probe correction for the electromagnetic fields satisfying Maxwell's equations

$$\nabla \times \bar{H} = \epsilon \frac{\partial}{\partial t} \bar{E}, \quad \nabla \times \bar{E} = -\mu \frac{\partial}{\partial t} \bar{H}, \quad \nabla \cdot \bar{E} = 0, \quad \nabla \cdot \bar{H} = 0 \tag{2.15}$$

in the free-space region $r > r_s$.

The time-domain electric field $\bar{E}(\bar{r},t)$ is determined from the corresponding frequency-domain field $\bar{E}_\omega(\bar{r})$ (and vice versa) by the Fourier transform identities

$$\bar{E}(\bar{r},t) = \int_{-\infty}^{+\infty} \bar{E}_\omega(\bar{r}) e^{-i\omega t} d\omega, \quad \bar{E}_\omega(\bar{r}) = \frac{1}{2\pi}\int_{-\infty}^{+\infty} \bar{E}(\bar{r},t) e^{i\omega t} dt \tag{2.16}$$

and outside the source region the frequency-domain field $\bar{E}_\omega$ satisfies the homogeneous vector Helmholtz equation, has zero divergence, and satisfies the radiation conditions [18, p.58].

In this section we will derive a time-domain near-field formula that is based on the following frequency-domain expansion of the tangential electric field into spherical vector-wave functions [25, pp.414-420], [12] [2]

$$\bar{E}_\omega^{tan}(\bar{r}) = \sum_{n=1}^{\infty}\sum_{m=-n}^{n}\left[\frac{h_n^{(1)}(kr)}{h_n^{(1)}(ka)} A_{nm}^\omega \bar{M}_{nm}(\theta,\phi) + \frac{g_n(kr)}{g_n(ka)} B_{nm}^\omega \bar{N}_{nm}(\theta,\phi)\right] \tag{2.17}$$

where $r \geq a$.

The tangential electric field is defined by $\bar{E}^{tan} = \bar{E} - \hat{r}\cdot\bar{E}\hat{r}$, the function $g_n(z)$ is given by

$$g_n(z) = \frac{1}{z}\frac{\partial}{\partial z}\left(z h_n^{(1)}(z)\right), \tag{2.18}$$

and transverse vector-wave functions

$$\bar{M}_{nm}(\theta,\phi) = \sqrt{\frac{2n+1}{4\pi n(n+1)}\frac{(n-m)!}{(n+m)!}}\left[\hat{\theta}\frac{im}{\sin\theta}P_n^m(\cos\theta) - \hat{\phi}\frac{\partial}{\partial\theta}P_n^m(\cos\theta)\right] e^{im\phi} \tag{2.19}$$

$$\bar{N}_{nm}(\theta,\phi) = \sqrt{\frac{2n+1}{4\pi n(n+1)}\frac{(n-m)!}{(n+m)!}}\left[\hat{\theta}\frac{\partial}{\partial\theta}P_n^m(\cos\theta) + \hat{\phi}\frac{im}{\sin\theta}P_n^m(\cos\theta)\right] e^{im\phi} \tag{2.20}$$

satisfy $\bar{M}_{nm} = \bar{N}_{nm} \times \hat{r}$ and the orthogonality relations

$$\int_0^{2\pi}\int_0^{\pi} \bar{M}_{nm}\cdot\bar{N}_{qs}^* \sin\theta d\theta d\phi = 0 \tag{2.21}$$

$$\int_0^{2\pi}\int_0^{\pi} \bar{M}_{nm}\cdot\bar{M}_{qs}^* \sin\theta d\theta d\phi = \int_0^{2\pi}\int_0^{\pi} \bar{N}_{nm}\cdot\bar{N}_{qs}^* \sin\theta d\theta d\phi = \delta_{nq}\delta_{ms} \tag{2.22}$$

where δ_{nq} is the Kronecker delta. The definitions (2.19) and (2.20) of the transverse vector-wave functions are valid for both positive and negative m and the associated Legendre functions $P_n^m(z)$ are given in [20]. With

[2] Although the expansion (2.17) was derived for $\omega > 0$ it is easily shown to be valid for all real ω. This is proven by using the same type of argument that was used to prove that (2.3) is valid for all real ω.

these definitions the expansion coefficients A^{ω}_{nm} and B^{ω}_{nm} in (2.17) are given in terms of the tangential electric field on the scan sphere by the simple integrations

$$A^{\omega}_{nm} = \int_0^{2\pi}\int_0^{\pi} \bar{E}_{\omega}(a\hat{r}) \cdot \bar{M}^*_{nm}(\theta,\phi)\sin\theta d\theta d\phi \tag{2.23}$$

and

$$B^{\omega}_{nm} = \int_0^{2\pi}\int_0^{\pi} \bar{E}_{\omega}(a\hat{r}) \cdot \bar{N}^*_{nm}(\theta,\phi)\sin\theta d\theta d\phi. \tag{2.24}$$

Taking the inverse Fourier transform of (2.17) shows that the tangential time-domain electric field outside the scan sphere is

$$\begin{aligned}\bar{E}^{tan}(\bar{r},t) = \frac{1}{2\pi}\sum_{n=1}^{\infty}\sum_{m=-n}^{n}\Bigg[\int_{-\infty}^{+\infty} & F_n(r,t')A_{nm}(t-t')dt'\bar{M}_{nm}(\theta,\phi) \\ + \int_{-\infty}^{+\infty} & G_n(r,t')B_{nm}(t-t')dt'\bar{N}_{nm}(\theta,\phi)\Bigg], \quad r \geq a\end{aligned} \tag{2.25}$$

where we have used the convolution theorem. From (2.23) and (2.24) it follows that the time-domain expansion coefficients are

$$A_{nm}(t) = \int_0^{2\pi}\int_0^{\pi} \bar{E}(a\hat{r},t) \cdot \bar{M}^*_{nm}(\theta,\phi)\sin\theta d\theta d\phi \tag{2.26}$$

and

$$B_{nm}(t) = \int_0^{2\pi}\int_0^{\pi} \bar{E}(a\hat{r},t) \cdot \bar{N}^*_{nm}(\theta,\phi)\sin\theta d\theta d\phi. \tag{2.27}$$

The function $F_n(r,t)$ is defined in (2.8) and was found to be given by the expression (2.12). Similarly the function $G_n(r,t)$ is defined by

$$G_n(r,t) = \int_{-\infty}^{+\infty} \frac{g_n(kr)}{g_n(ka)} e^{-i\omega t} d\omega \tag{2.28}$$

with g_n given by (2.18). To determine an explicit expression for G_n note that the results of [21] and [22, app. B] show that the function $g_n(z)$ $(n \geq 1)$ has $n+1$ simple zeros located symmetrically around the imaginary axis inside the circle of radius n in the half plane $\mathrm{Im}(z) < 0$. In [22, app. B] an asymptotic formula is presented that determines the zeros of $g_n(z)$ for large n. Denoting the corresponding zeros in the complex frequency plane by ω'_{ns} we have

$$g_n(a\omega'_{ns}/c) = 0, \quad s = 1,2,3,\ldots,n,n+1 \tag{2.29}$$

with $|a\omega'_{ns}/c| < n$ and $\mathrm{Im}(\omega'_{ns}) < 0$. The zeros ω'_{ns} lie in a pattern similar to that in Figure 2. Using the residue theorem and the asymptotic form of the function $g_n(z)$ it is found that

$$G_n(r,t) = \frac{2\pi a}{r}\delta(t-r/c+a/c) - u(t-r/c+a/c)\frac{2\pi ic}{a}\sum_{s=1}^{n+1}\frac{g_n(\omega'_{ns}r/c)}{g'_n(\omega'_{ns}a/c)}e^{-i\omega'_{ns}t} \tag{2.30}$$

where $u(t)$ is the unit step function and the contour C'_n encloses all the zeros ω'_{ns}. The function $g'_n(z)$ is the derivative of $g_n(z)$ with respect to z.

Having determined expressions for both F_n and G_n we can now calculate the convolution integrals in the time-domain expansion (2.25) and find that for $r \geq a$

$$\begin{aligned}
\bar{E}^{tan}(\bar{r},t) = \sum_{n=1}^{\infty}\Bigg[\frac{a}{r}\sum_{m=-n}^{n} & A_{nm}(t-r/c+a/c)\bar{M}_{nm}(\theta,\phi) \\
& + B_{nm}(t-r/c+a/c)\bar{N}_{nm}(\theta,\phi) \\
- \frac{ic}{a}\sum_{s=1}^{n}\frac{h_n^{(1)}(\omega_{ns}r/c)}{h_n^{(1)\prime}(\omega_{ns}a/c)}\sum_{m=-n}^{n}\int_{t_0}^{t-r/c+a/c} & A_{nm}(t')e^{-i\omega_{ns}(t-t')}dt'\,\bar{M}_{nm}(\theta,\phi) \\
- \frac{ic}{a}\sum_{s=1}^{n+1}\frac{g_n(\omega'_{ns}r/c)}{g'_n(\omega'_{ns}a/c)}\sum_{m=-n}^{n}\int_{t_0}^{t-r/c+a/c} & B_{nm}(t')e^{-i\omega'_{ns}(t-t')}dt'\,\bar{N}_{nm}(\theta,\phi)\Bigg]
\end{aligned} \tag{2.31}$$

where the zeros ω_{ns} and ω'_{ns} are given by (2.11) and (2.29), respectively, and t_0 is determined such that all fields are zero in the region $r \geq a$ for $t < t_0$. Because the zeros ω_{ns} and ω'_{ns} are located symmetrically with respect to the imaginary axis, one can show that the right side of (2.31) indeed is real. The time-domain spherical near-field formula (2.31) gives the tangential electric field outside the scan sphere in terms of the expansion coefficients (2.26) and (2.27) that are determined from the tangential electric field on the scan sphere.

The electric far-field pattern $\bar{\mathcal{F}}(\theta,\phi,t)$ is defined such that the far electric field is $\bar{E}(\bar{r},t) \sim \bar{\mathcal{F}}(\theta,\phi,t-r/c)/r$ for $r \to \infty$. Substituting the large-argument approximations for $h_n^{(1)}(z)$ and $g_n(z)$ into (2.31) shows that this far-field pattern is

$$\begin{aligned}
\bar{\mathcal{F}}(\theta,\phi,t) = a\sum_{n=1}^{\infty}\sum_{m=-n}^{n}\Bigg[& A_{nm}(t+a/c)\bar{M}_{nm}(\theta,\phi) + B_{nm}(t+a/c)\bar{N}_{nm}(\theta,\phi) \\
& -\frac{c^2(-i)^n}{a^2}\sum_{s=1}^{n}\frac{e^{-i\omega_{ns}t}}{\omega_{ns}h_n^{(1)\prime}(\omega_{ns}a/c)}\int_{t_0}^{t+a/c} A_{nm}(t')e^{i\omega_{ns}t'}dt'\,\bar{M}_{nm}(\theta,\phi)
\end{aligned}$$

$$+\frac{c^2(-i)^{n+1}}{a^2}\sum_{s=1}^{n+1}\frac{e^{-i\omega'_{ns}t}}{\omega'_{ns}g'_n(\omega'_{ns}a/c)}\int_{t_0}^{t+a/c}B_{nm}(t')e^{i\omega'_{ns}t'}dt'\bar{N}_{nm}(\theta,\phi)\Bigg]. \tag{2.32}$$

We have now derived the time-domain analogs to the frequency-domain spherical near-field formulas that have been used extensively for determining frequency-domain far-fields from measured frequency-domain near fields [10]. The time-domain formulas (2.31) and (2.32) could also have been derived from the dyadic version of Green's second identity using the Dirichlet dyadic Green's function for the perfectly conducting sphere. Thus, it is not surprising that the complex resonant frequencies for the perfectly conducting sphere [26] occur in these spherical near-field formulas.

3. Formulas with probe correction. To use the formulas of Section 2 one has to know the actual fields on the scan sphere. In many practical measurement situations, however, the measurements on the scan sphere are taken with a nonideal probe. The output of such a probe is not related in a simple manner to the illuminating field, and thus in such situations, it is not immediately possible to use the formulas of Section 2 to compute the fields outside the scan sphere.

In order to compute the fields outside the scan sphere from measurements obtained with a nonideal probe one needs probe-corrected formulas that express the desired fields in terms of the output of the probe. In this section we shall employ the spherical expansions of Section 2 along with the operator approach of Yaghjian and Wittmann [11], [12] to derive the probe-corrected spherical near-field formulas for both acoustic and electromagnetic fields in the time domain. It is assumed throughout that multiple interactions between the probe and the source region are negligible. That is, the reflection in the source region, of the field radiated by the induced sources on the probe, does not affect the field measured by the probe.

In Section 3.1 we derive probe-corrected spherical near-field formulas for the acoustic field that are based on the eigenfunction expansion of Section 2.1. Similarly, in Section 3.2 we derive probe-corrected near-field formulas for the electromagnetic field based on the vector-wave expansion in Section 2.2. As in the previous section, the arbitrary finite source region is located inside the minimum sphere of radius r_s and we measure the field on the scan sphere of radius $a > r_s$.

3.1. Acoustic fields. We begin with a presentation of the linear differential operator formula that expresses the output of the probe in terms of the illuminating field and its spatial derivatives at a reference point of the probe. The linear differential operator formula for acoustic fields was first derived by Yaghjian [11] and a detailed rederivation can be found in [22] and [24].

The probe coordinate system (x, y, z) is fixed with respect to the probe such that the axis of the probe is parallel to the z axis as shown in Figure

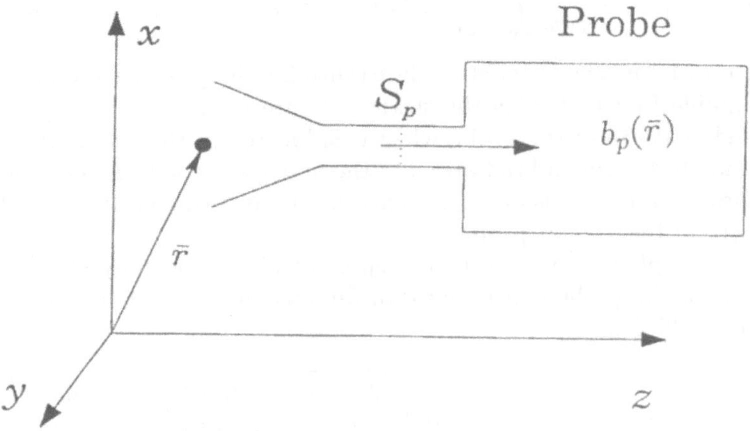

FIG. 3. *The probe coordinate system.*

3. We denote the location of some reference point of the probe by $\bar{r}$. The frequency-domain receiving characteristic of the probe is denoted by $R^p_\omega(k_x, k_y)$ and is defined to be the output of the probe, when its reference point is located at the origin of the probe coordinate system in Figure 3, and the incident field is the frequency-domain plane wave $(2\pi)^{-1}e^{i\bar{k}\cdot\bar{r}}$ [14, sec.2.1], [17]. The propagation vector of this incident plane wave is $\bar{k} = k_x\hat{x} + k_y\hat{y} + \gamma\hat{z}$ with $\gamma = |\sqrt{k^2 - k_x^2 - k_y^2}|$ for $k_x^2 + k_y^2 < k^2$ and $\gamma = i|\sqrt{k_x^2 + k_y^2 - k^2}|$ for $k_x^2 + k_y^2 > k^2$.

To derive the linear differential operator formula one begins by noting that the receiving characteristic $R^p_\omega(k_x, k_y)$ is an analytic function of the variables (k_x, k_y) and may be expanded in a series of spherical harmonics [11], [12], and [27]. This expansion is written as

$$R^p_\omega(k_x, k_y) = \sum_{n=0}^{\infty} \sum_{m=-n}^{n} a^\omega_{nm} \mathcal{Y}_{nm}(k_x, k_y) \tag{3.1}$$

where a^ω_{nm} are the probe receiving coefficients and $\mathcal{Y}_{nm}(k_x, k_y)$ are analytical continuations of the spherical harmonics [22], [24]. Assuming that the probe is a reciprocal electroacoustic transducer [14], [17], [28] fed by an electromagnetic waveguide in which only a single mode is propagating, one may show that the probe receiving coefficients are given by [22], [24]

$$(3.2) \qquad a_{nm}^{\omega} = \frac{i}{\omega \eta_0 \rho_0} \int_0^{2\pi} \int_0^{\pi} \mathcal{F}_{\omega}^{p}(\pi - \theta, \pi + \phi) Y_{nm}^{*}(\theta, \phi) \sin\theta d\theta d\phi$$

where η_0 is the characteristic admittance for the propagating mode in the waveguide feed and ρ_0 is the static pressure of the fluid. $\mathcal{F}_{\omega}^{p}(\theta, \phi)$ is the far-field pattern of the probe when it is located at the origin of the probe coordinate system and radiates into the half space $z < 0$. If the probe is not reciprocal, the expansion coefficients can be determined from the far-field pattern of the adjoint probe.

Using plane-wave spectrum representations one may show that the output of the probe can be written directly in terms of the incident field as [11], [22], [24]

$$b_{p\omega}(\bar{r}) = 2\pi \sum_{n=0}^{\infty} \sum_{m=-n}^{n} a_{nm}^{\omega} \sqrt{\frac{2n+1}{4\pi} \frac{(n-|m|)!}{(n+|m|)!}} \left(\frac{\operatorname{sgn}(-m)}{ik} \right)^{|m|}$$

$$(3.3) \qquad \cdot P_n^{(|m|)} \left(\frac{1}{ik} \frac{\partial}{\partial z} \right) \left(\frac{\partial}{\partial x} + i \operatorname{sgn}(m) \frac{\partial}{\partial y} \right)^{|m|} \Phi_{\omega}(\bar{r})$$

where $P_n^{(|m|)}$ denotes the $|m|$'th derivative of the Legendre polynomial. This is the linear differential operator expression that gives the output of the probe in terms of the probe receiving coefficients and spatial derivatives of the incident field at the reference point of the probe.

Following [11], [29] we assume that the probe is axis symmetric, that is, its receiving characteristic is independent of the angle k_{ϕ} when the variables (k_x, k_y) are written in terms of spherical coordinates. This means that the probe receiving coefficients satisfy $a_{nm}^{\omega} = 0$ for $m \neq 0$ and the expression (3.3) for the output of the probe reduces to

$$(3.4) \qquad b_{p\omega}(\bar{r}) = \sum_{n=0}^{\infty} a_n^{\omega} P_n \left(\frac{1}{ik} \frac{\partial}{\partial z} \right) \Phi_{\omega}(\bar{r})$$

where $a_n^{\omega} \equiv 2\pi \sqrt{\frac{2n+1}{4\pi}} a_{n0}^{\omega}$ and $P_n(z)$ is the Legendre polynomial. Because we have assumed that the probe is axis symmetric only derivatives of the incident field with respect to the axial direction z are needed to compute the output of the probe. Now that we have determined the output of the probe in the probe coordinate system shown in Figure 3 we proceed to derive the probe-corrected spherical near-field formulas.

When the probe is used to measure the field of the source region on the scan sphere in Figure 1, the origin of the probe coordinate system is always located at the scan point $a\hat{r}$ and the axial direction of the probe is always parallel to $\hat{r}$. Consequently, (3.4) shows that in the coordinate system (r, θ, ϕ), in which the scan sphere is given by the equation $r = a$, the output of the probe is

$$b_{p\omega}(a\hat{r}) = \sum_{n=0}^{\infty} a_n^{\omega} P_n\left(\frac{1}{ik}\frac{\partial}{\partial r}\right) \Phi_\omega(r\hat{r})\bigg|_{r=a} . \tag{3.5}$$

Inserting the expansion (2.3) for the field of the source region into the operator expression (3.5) for the output of the probe shows that

$$b_{p\omega}(a\hat{r}) = \sum_{n=0}^{\infty}\sum_{q=0}^{\infty} a_q^{\omega} P_q\left(\frac{1}{ik}\frac{\partial}{\partial r}\right) \frac{h_n^{(1)}(kr)}{h_n^{(1)}(ka)} \sum_{m=-n}^{n} C_{nm}^{\omega} Y_{nm}(\theta,\phi)\bigg|_{r=a} . \tag{3.6}$$

Employing the orthogonality relations for the spherical harmonics and defining the new expansion coefficients $C_{nm}^{p\omega}$ to be simply the double integral

$$C_{nm}^{p\omega} = \int_0^{2\pi}\int_0^{\pi} b_{p\omega}(a\hat{r}) Y_{nm}^*(\theta,\phi)\sin\theta d\theta d\phi, \tag{3.7}$$

where the superscript p indicates that these coefficients are found from the output of the probe, it is found from (2.3) and (3.6) that

$$\Phi_\omega(\bar{r}) = \sum_{n=0}^{\infty} \frac{h_n^{(1)}(kr)}{\sum_{q=0}^{\infty} a_q^{\omega} P_q\left(\frac{1}{ik}\frac{\partial}{\partial r}\right) h_n^{(1)}(kr)\bigg|_{r=a}} \sum_{m=-n}^{n} C_{nm}^{p\omega} Y_{nm}(\theta,\phi), \quad r \geq a. \tag{3.8}$$

This is the final expression that gives the frequency-domain acoustic field outside the scan sphere in terms of the output of the probe and its receiving coefficients a_n^{ω}.

Taking the inverse Fourier transform of (3.8) and using the convolution rule shows that the time-domain acoustic field outside the scan sphere is given by the expression

$$\Phi(\bar{r},t) = \frac{1}{2\pi}\sum_{n=0}^{\infty}\sum_{m=-n}^{n}\int_{-\infty}^{+\infty} L_n(r,t') C_{nm}^{p}(t-t')dt' Y_{nm}(\theta,\phi), \quad r \geq a \tag{3.9}$$

where the time functions $L_n(r,t)$ are defined by

$$L_n(r,t) = \int_{-\infty}^{+\infty} \frac{h_n^{(1)}(kr)}{\sum_{q=0}^{\infty} a_q^{\omega} P_q\left(\frac{1}{ik}\frac{\partial}{\partial r}\right) h_n^{(1)}(kr)\bigg|_{r=a}} e^{-i\omega t} d\omega, \tag{3.10}$$

and the time-domain expansion coefficients $C_{nm}^{p}(t)$ are given by

$$C_{nm}^{p}(t) = \int_0^{2\pi}\int_0^{\pi} b_p(a\hat{r},t) Y_{nm}^*(\theta,\phi)\sin\theta d\theta d\phi. \tag{3.11}$$

Note that the form of the probe-corrected formula (3.9) is the same as that of the non-probe-corrected formula (2.9). In the probe-corrected formula (3.9) the time-domain expansion coefficients are determined by integrating the output of the probe over the scan sphere, whereas in the non-probe-corrected formula (2.9) these coefficients are determined by integrating the acoustic field over the scan sphere. The time functions $L_n(r,t)$ in the probe-corrected formula (3.9) are given in terms of the probe receiving coefficients a_n^ω and the spherical Hankel function, whereas the time functions $F_n(r,t)$ in the non-probe-corrected formula (2.9) are given in terms of spherical Hankel functions only. Once the probe receiving coefficients a_n^ω are known, the singularities (poles) of the integrand in (3.10) may be determined. After this is done, the functions $L_n(r,t)$ can be computed explicitly with the help of complex function theory and the technique that was used to calculate $F_n(r,t)$ in Section 2.1.

To illustrate the use of the general probe-corrected formulas (3.9)-(3.11) we shall consider a special time-derivative probe for which the functions $L_n(r,t)$ can be computed explicitly. Time-derivative probes have been discussed in [14, 3.1.2], [17] and can be characterized in the frequency-domain as follows. The output of such a probe, due to the incoming plane wave $(2\pi)^{-1}e^{i(k_x x+k_y y+\gamma z)}$ is equal to ω multiplied by a function that depends only on the direction of incidence of the plane wave. According to [14, eq.(3.16)], [17] the receiving characteristic of this probe is given by

$$R_p^\omega = -\frac{i\omega Q_p}{2\pi} \tag{3.12}$$

where the function Q_p is *independent* of the frequency. In this section we consider the special time-derivative probe for which the function Q_p is equal to $\cos\theta$ for the propagating modes. The angle θ denotes the direction of propagation of the incident plane wave when the probe is located in the probe coordinate system in Figure 3. Using analytic continuation it is found that $\omega Q_p = c\gamma$ for both propagating and evanescent modes. Inserting this expression into (3.1) it is found that the probe receiving coefficients are given by $a_{1,0}^\omega = -ick(3\pi)^{-1/2}$ and $a_{nm}^\omega = 0$ for $(n,m) \neq (1,0)$. This means that the coefficients a_n^ω defining the function $L_n(r,t)$ in (3.10) are given by $a_1^\omega = -ick$ and $a_n^\omega = 0$ for $n \neq 1$. Inserting these values of a_n^ω into (3.10) and using the residue theorem shows that $L_n(r,t)$ can be expressed by the residue series

$$L_n(r,t) = 2\pi i u(t-r/c+a/c)\sum_{s=1}^{n+1} \frac{h_n^{(1)}(\underline{\omega}_{ns}r/c)}{(\underline{\omega}_{ns}a/c)h_n^{(1)''}(\underline{\omega}_{ns}a/c)} e^{-i\underline{\omega}_{ns}t} \tag{3.13}$$

where $u(t)$ is the unit-step function, $\underline{\omega}_{ns}$ are the solutions to the equation $h_n^{(1)'}(a\underline{\omega}_{ns}/c) = 0$, and $h_n^{(1)''}(z)$ is the second derivative of the spherical Hankel function with respect to its argument. Olver [21] has shown that the zeros $\underline{\omega}_{ns}$ lie in a pattern similar to that in Figure 2. Having derived the closed-form expression for $L_n(r,t)$ we insert this expression into the probe-corrected formula (3.9) and find that for $r \geq a$

$$\Phi(\bar{r},t) = i\sum_{n=0}^{\infty}\sum_{m=-n}^{n}\sum_{s=1}^{n+1}\frac{h_n^{(1)}(\underline{\omega}_{ns}r/c)}{(\underline{\omega}_{ns}a/c)h_n^{(1)''}(\underline{\omega}_{ns}a/c)}\int_{t_0}^{t-r/c+a/c} C_{nm}^{p}(t')e^{-i\underline{\omega}_{ns}(t-t')}dt'Y_{nm}(\theta,\phi) \tag{3.14}$$

where we have used the fact that all fields in the region $r \geq a$ are zero for $t < t_0$. Along with the expression (3.11) for the coefficients $C_{nm}^{p}(t)$, equation (3.14) expresses the field outside the scan sphere in terms of the output of the special time-derivative probe on the scan sphere.

3.2. Electromagnetic fields. As in the previous section we begin with a description of the probe when it is located in the probe coordinate system with its axis parallel to the z axis as shown in Figure 3. The vector receiving characteristic of the probe $\bar{R}_{\omega}^{p}(k_x,k_y)$ is defined such that the scalar output of the probe is given by $b_p = \bar{R}_{\omega}^{p}(k_x,k_y)\cdot\bar{e}$, when the reference point of the probe is located at the origin of coordinate system in Figure 3, and the incident plane wave is given by $(2\pi)^{-1}\bar{e}e^{i\bar{k}\cdot\bar{r}}$ with $\bar{e}\cdot\bar{k}=0$.

We expand the vector receiving characteristic of the probe in terms of spherical vector-wave functions as

$$\bar{R}_{\omega}^{p}(k_x,k_y) = \sum_{n=1}^{\infty}\sum_{m=-n}^{n}\left[a_{nm}^{\omega}\bar{\mathcal{M}}_{nm}(k_x,k_y) + b_{nm}^{\omega}\bar{\mathcal{N}}_{nm}(k_x,k_y)\right] \tag{3.15}$$

where a_{nm}^{ω} and b_{nm}^{ω} are the probe receiving coefficients and $\bar{\mathcal{M}}_{nm}(k_x,k_y)$ and $\bar{\mathcal{N}}_{nm}(k_x,k_y)$ are analytical continuations of the spherical vector-wave functions [22]. If the probe is reciprocal and is fed by a waveguide supporting just one propagating mode with a characteristic admittance η_0, the receiving coefficients are [22, sec.3.2], [23]

$$a_{nm}^{\omega} = \frac{i}{k\eta_0}\sqrt{\frac{\epsilon}{\mu}}\int_0^{2\pi}\int_0^{\pi}\bar{\mathcal{F}}_{\omega}^{p}(\pi-\theta,\pi+\phi)\cdot\bar{M}_{nm}^{*}(\theta,\phi)\sin\theta d\theta d\phi \tag{3.16}$$

and

$$b_{nm}^{\omega} = \frac{i}{k\eta_0}\sqrt{\frac{\epsilon}{\mu}}\int_0^{2\pi}\int_0^{\pi}\bar{\mathcal{F}}_{\omega}^{p}(\pi-\theta,\pi+\phi)\cdot\bar{N}_{nm}^{*}(\theta,\phi)\sin\theta d\theta d\phi \tag{3.17}$$

where $\bar{\mathcal{F}}_{\omega}^{p}(\theta,\phi)$ is the far-field pattern of the probe in the probe coordinate system when it radiates into the half space $z < 0$. If the probe is not reciprocal the probe receiving coefficients are given by (3.16) and (3.17) with the far-field pattern of the probe replaced by the far-field pattern of the adjoint probe.

The probes that are normally used for spherical near-field measurements have receiving characteristics that satisfy [12, eqs.(20)], [10, p.151]

$$a_{nm}^{\omega} = b_{nm}^{\omega} = 0, \text{ for } m \neq \pm 1, \text{ and } a_{n1}^{\omega} = -a_{n,-1}^{\omega} \equiv a_n^{\omega}, \quad b_{n1}^{\omega} = b_{n,-1}^{\omega} \equiv b_n^{\omega}. \tag{3.18}$$

With these values for the receiving coefficients one finds the following linear differential operator expression for the output of the probe [12], [22, sec.3.2], [23]

$$b_{p\omega}(\bar{r}) = -\sqrt{3\pi}\sum_{n=1}^{\infty}\left[\left\{a_n^{\omega}\beta_{n1}\left(\frac{1}{ik}\frac{\partial}{\partial z}\right) + ib_n^{\omega}\alpha_{n1}\left(\frac{1}{ik}\frac{\partial}{\partial z}\right)\right\}\sqrt{\frac{\mu}{\epsilon}}\bar{H}_{\omega}(\bar{r})\cdot\hat{x}\right.$$

$$\left. + \left\{a_n^{\omega}\alpha_{n1}\left(\frac{1}{ik}\frac{\partial}{\partial z}\right) + ib_n^{\omega}\beta_{n1}\left(\frac{1}{ik}\frac{\partial}{\partial z}\right)\right\}\bar{E}_{\omega}(\bar{r})\cdot\hat{y}\right] \tag{3.19}$$

and we see that the output of this special probe is determined from z-derivatives of the x and y components of the magnetic and electric field, respectively. Furthermore, α_{n1} and β_{n1} are polynomials given by

$$\alpha_{n1}(z) = \sqrt{\frac{4n+2}{3n(n+1)}}\,P_n''(z) \tag{3.20}$$

and

$$\beta_{n1}(z) = \sqrt{\frac{4n+2}{3n(n+1)}}\,[zP_n''(z) + P_n'(z)] \tag{3.21}$$

where P_n' and P_n'' are the fist and second derivative of the Legendre polynomial, respectively.

To obtain a convenient form of the expression for the output of the probe, define the two operators

$$\mathcal{L}_{\omega}^{H}(z) = -\sqrt{3\pi}\sum_{n=1}^{\infty}\left\{a_n^{\omega}\beta_{n1}\left(\frac{1}{ik}\frac{\partial}{\partial z}\right) + ib_n^{\omega}\alpha_{n1}\left(\frac{1}{ik}\frac{\partial}{\partial z}\right)\right\} \tag{3.22}$$

and

$$\mathcal{L}_{\omega}^{E}(z) = -\sqrt{3\pi}\sum_{n=1}^{\infty}\left\{a_n^{\omega}\alpha_{n1}\left(\frac{1}{ik}\frac{\partial}{\partial z}\right) + ib_n^{\omega}\beta_{n1}\left(\frac{1}{ik}\frac{\partial}{\partial z}\right)\right\} \tag{3.23}$$

so that the output of the probe can be written as

$$b_{p\omega}(\bar{r}) = \sqrt{\frac{\mu}{\epsilon}}\mathcal{L}_{\omega}^{H}(z)\bar{H}_{\omega}(\bar{r})\cdot\hat{x} + \mathcal{L}_{\omega}^{E}(z)\bar{E}_{\omega}(\bar{r})\cdot\hat{y}. \tag{3.24}$$

This is the final linear differential operator expression for the output of the probe, given in terms of z-derivatives of the electromagnetic fields at the reference point of the probe.

Having derived the linear differential operator formula for the output of the probe, we can now derive the probe-corrected formulas for the field outside the scan sphere. When the probe is used to measure the field on the scan sphere, the origin of the probe coordinate system is located at the scan point $a\hat{r}$ and the axial direction of the probe (which is also the z axis of the probe coordinate system) is always parallel to $\hat{r}$. With the x axis of the probe coordinate system shown in Figure 3 parallel to the spherical θ unit vector, the output of the probe is denoted by $b_{p\omega}^{\psi=0}(\bar{r})$ and is found from (3.24) to be given by

$$b_{p\omega}^{\psi=0}(\bar{r}) = \sqrt{\frac{\mu}{\epsilon}} \mathcal{L}_\omega^H(r)\bar{H}_\omega(\bar{r}) \cdot \hat{\theta} + \mathcal{L}_\omega^E(r)\bar{E}_\omega(\bar{r}) \cdot \hat{\phi} \tag{3.25}$$

where the operators $\mathcal{L}_\omega^H$ and $\mathcal{L}_\omega^E$ are expressed in (3.22) and (3.23) in terms of the receiving coefficients of the probe. Similarly, when the x axis of the probe coordinate system is parallel to the spherical ϕ unit vector, the output of the probe is denoted by $b_{p\omega}^{\psi=90}(\bar{r})$ and is found from (3.24) to be given by

$$b_{p\omega}^{\psi=90}(\bar{r}) = \sqrt{\frac{\mu}{\epsilon}} \mathcal{L}_\omega^H(r)\bar{H}_\omega(\bar{r}) \cdot \hat{\phi} - \mathcal{L}_\omega^E(r)\bar{E}_\omega(\bar{r}) \cdot \hat{\theta}. \tag{3.26}$$

We can thus define a vector output of the probe by $\bar{b}_{p\omega} = \hat{\theta} b_{p\omega}^{\psi=0} + \hat{\phi} b_{p\omega}^{\psi=90}$ and have [12, eq. (23)]

$$\bar{b}_{p\omega}(\bar{r}) = \sqrt{\frac{\mu}{\epsilon}} \mathcal{L}_\omega^H(r)\bar{H}_\omega^{tan}(\bar{r}) + \mathcal{L}_\omega^E(r)\bar{E}_\omega^{tan}(\bar{r}) \times \hat{r} \tag{3.27}$$

where the tangential magnetic field is $\bar{H}^{tan} = \bar{H} - \hat{r} \cdot \bar{H}\hat{r}$ and the tangential electric field is defined similarly.

From the vector-wave function expansion (2.17) of the tangential electric field it is easy to determine expansions for $\sqrt{\frac{\mu}{\epsilon}}\bar{H}_\omega^{tan}(\bar{r})$ and $\bar{E}_\omega^{tan}(\bar{r}) \times \hat{r}$. Inserting these expansions into the expression (3.27) for the vector output of the probe shows that when the probe is illuminated by the field generated by the source region in Figure 1 its output is

$$\bar{b}_{p\omega}(\bar{r}) = \sum_{n=1}^{\infty} \sum_{m=-n}^{n} \left[\left\{ -i\mathcal{L}_\omega^H(r) g_n(kr) - \mathcal{L}_\omega^E(r) h_n^{(1)}(kr) \right\} \frac{A_{nm}^\omega}{h_n^{(1)}(ka)} \bar{N}_{nm}(\theta, \phi) \right.$$
$$\left. + \left\{ -i\mathcal{L}_\omega^H(r) h_n^{(1)}(kr) + \mathcal{L}_\omega^E(r) g_n(kr) \right\} \frac{B_{nm}^\omega}{g_n(ka)} \bar{M}_{nm}(\theta, \phi) \right]. \tag{3.28}$$

It is now a simple matter to employ the orthogonality relations for the spherical vector-wave functions to determine expressions for the expansion coefficients A_{nm}^ω and B_{nm}^ω. Once these coefficients are determined, equation (2.17) gives the frequency-domain tangential electric field outside the scan sphere. Employing the Fourier transform then gives the desired probe-corrected spherical near-field formulas for time-domain electromagnetic fields [22, sec.3.2], [23]:

$$\bar{E}^{tan}(\bar{r},t) = \frac{1}{2\pi}\sum_{n=1}^{\infty}\sum_{m=-n}^{n}\left[\int_{-\infty}^{+\infty} F_n^p(r,t')A_{nm}^p(t-t')dt'\bar{M}_{nm}(\theta,\phi)\right.$$

$$\left. + \int_{-\infty}^{+\infty} G_n^p(r,t')B_{nm}^p(t-t')dt'\bar{N}_{nm}(\theta,\phi)\right], \quad r \geq a \tag{3.29}$$

where the time functions F_n^p and G_n^p are

$$F_n^p(r,t) = \int_{-\infty}^{+\infty} \frac{h_n^{(1)}(kr)e^{-i\omega t}}{\left[-i\mathcal{L}_\omega^H(r)g_n(kr) - \mathcal{L}_\omega^E(r)h_n^{(1)}(kr)\right]_{r=a}} d\omega \tag{3.30}$$

and

$$G_n^p(r,t) = \int_{-\infty}^{+\infty} \frac{g_n(kr)e^{-i\omega t}}{\left[-i\mathcal{L}_\omega^H(r)h_n^{(1)}(kr) + \mathcal{L}_\omega^E(r)g_n(kr)\right]_{r=a}} d\omega. \tag{3.31}$$

The time-domain coefficients $A_{nm}^p(t)$ and $B_{nm}^p(t)$ are simply

$$A_{nm}^p(t) = \int_0^{2\pi}\int_0^{\pi} \bar{b}_p(a\hat{r},t)\cdot\bar{N}_{nm}^*(\theta,\phi)\sin\theta d\theta d\phi \tag{3.32}$$

and

$$B_{nm}^p(t) = \int_0^{2\pi}\int_0^{\pi} \bar{b}_p(a\hat{r},t)\cdot\bar{M}_{nm}^*(\theta,\phi)\sin\theta d\theta d\phi. \tag{3.33}$$

Equation (3.29) is the probe-corrected spherical near-field formula in the time domain that is based on the linear differential operator description of the probe. To use this near-field formula one must first use (3.22) and (3.23) to determine the operators $\mathcal{L}_\omega^H$ and $\mathcal{L}_\omega^E$ from the probe receiving coefficients a_{nm}^ω and b_{nm}^ω in (3.16) and (3.17). Having determined the operators one must then compute the Fourier integrals (3.30) and (3.31) either numerically or with the help of complex function theory. Then the vector output of the probe is obtained on the scan sphere and the time-domain coefficients $A_{nm}^p(t)$ and $B_{nm}^p(t)$ are computed from the simple integrations in (3.32) and (3.33). Finally, the time functions and the time-domain coefficients are inserted into (3.29) to get an expression for the field outside the scan sphere.

4. Sampling theorems and numerical field calculations. The formulas derived so far for the field outside the scan sphere involve infinite summations and require that the field at all points on the scan sphere is known. We shall now show that the eigenfunction expansions can be truncated when the time-domain fields are bandlimited. And from these truncated eigenfunction expansions we derive formulas that give the expansion coefficients in terms of sampled near-field data. For the formulas without probe correction the sampled near-field data consist of sampled values in space-time of the field on the scan sphere; whereas for the probe-corrected formulas the sampled near-field data consist of sampled values in space-time of the output of the probe on the scan sphere. Also presented in this section is a numerical example in which the field of an acoustic point source with Gaussian time dependence is computed with a truncated eigenfunction expansion from sampled near-field data.

4.1. Sampling theorems for acoustic fields. We begin with the formulas without probe correction and assume that the acoustic field on the scan sphere is bandlimited so that $\Phi_\omega(\bar{r}) = \frac{1}{2\pi}\int_{-\infty}^{+\infty} \Phi(\bar{r},t)e^{i\omega t}dt$ can be set equal to zero for $|\omega| > \omega_{max}$. Then the frequency-domain coefficients (2.4) are zero for $|\omega| > \omega_{max}$ and it follows from [10, p.17] that all terms with $n > N$ in the frequency-domain eigenfunction expansion (2.3) can be set equal to zero. For most antennas, the integer N is determined by the radius r_s of the minimum sphere and the maximum wave number $k_{max} = \omega_{max}/c$ by the equation

$$N = [k_{max} r_s] + n_1 \tag{4.1}$$

where $[k_{max} r_s]$ is the integral part of $k_{max} r_s$ and n_1 is an integer depending on the location of the coordinate system, the source, the observation point, and the accuracy required [10, p.17]. In Section 4.3 we show a numerical example that illustrates the effect n_1 has on the accuracy of the computed fields.

Taking the inverse Fourier transform of the truncated version of the frequency-domain expansion (2.3) and using the same technique that led from (2.3) to (2.13) it follows that the time-domain acoustic field is given by the truncated expansion

$$\Phi(\bar{r},t) = \sum_{n=0}^{N} \left\{ \frac{a}{r} \sum_{m=-n}^{n} C_{nm}(t - r/c + a/c) Y_{nm}(\theta,\phi) \right.$$

$$\left. -\frac{ic}{a} \sum_{s=1}^{n} \frac{h_n^{(1)}(\omega_{ns} r/c)}{h_n^{(1)'}(\omega_{ns} a/c)} \sum_{m=-n}^{n} \int_{t_0}^{t-r/c+a/c} C_{nm}(t') e^{-i\omega_{ns}(t-t')} dt' Y_{nm}(\theta,\phi) \right\}, \tag{4.2}$$

where all fields on the scan sphere have been assumed zero for $t < t_0$. [The time-domain far-field pattern is similarly given by (2.14) with the upper limit of the summation over n changed from ∞ to N.]

We shall now use the truncated expansion (4.2) to show how the time-domain expansion coefficients $C_{nm}(t)$ given by (2.7) can be computed from sampled values of the field on the scan sphere. This will be done by first deriving a Fourier series expansion for the field on the scan sphere in which the Fourier coefficients are determined from sampled values of the field on the scan sphere. Then this Fourier series expansion is inserted into the expression (2.7) for the time-domain expansion coefficients to get the desired expression for the expansion coefficients in terms of sampled near-field data.

From the expression for the spherical harmonics (2.5) it is seen that with respect to ϕ the function $\Phi(\bar{r},t)$ in (4.2) is a spatially bandlimited function with $2N+1$ nonzero Fourier coefficients. Specifically, when the observation point is on the scan sphere we have

$$\Phi(a\hat{r}(\theta,\phi),t) = \sum_{m=-N}^{N} c_m(\theta,t)e^{im\phi} \tag{4.3}$$

where the Fourier coefficients $c_m(\theta,t)$ are given by

$$c_m(\theta,t) = \frac{1}{2\pi}\int_0^{2\pi} \Phi(a\hat{r}(\theta,\phi),t)e^{-im\phi}d\phi \tag{4.4}$$

and $\hat{r}(\theta,\phi) = \hat{x}\cos\theta\sin\phi + \hat{y}\sin\theta\sin\phi + \hat{z}\cos\theta$. From the standard sampling theorem for periodic spatially bandlimited functions [10, p.372] it follows that the integrals in (4.4) can be computed *exactly* from sampled values of Φ as

$$c_m(\theta,t) = \frac{1}{2N+1}\sum_{l_\phi=0}^{2N} \Phi(a\hat{r}(\theta,l_\phi\Delta\phi),t)e^{-iml_\phi\Delta\phi}, \quad m=-N,...,0,...N \tag{4.5}$$

where $\Delta\phi = 2\pi/(2N+1)$. [Here l_ϕ is an integer determining the value of the angle ϕ at the scan point by the equation $\phi = l_\phi\Delta\phi$.]

We will now show how the expansion coefficients $c_m(\theta,t)$ can be determined from sampled values of the field on the scan sphere. Begin by noting that from the expression (2.5) for the spherical harmonics it follows that the truncated expansion (4.2) shows that the function $c_m(\theta,t)$ can be expanded in terms of the functions $P_n^m(\cos\theta)$ with $n = |m|, |m|+1, ..., N$. Since $P_n^m(\cos\theta)$ can be expanded in terms of the complex exponentials $e^{iq\theta}$ with $q = -n, -n+1, ..., 0, ...n-1, n$ we find that

$$c_m(\theta,t) = \sum_{q=-N}^{N} d_{qm}(t)e^{iq\theta}, \quad 0 \le \theta \le \pi. \tag{4.6}$$

Because $c_m(\theta,t)$ is *not* periodic on the interval $0 \le \theta \le \pi$ we cannot immediately use the standard sampling theorem for periodic functions to determine the expansion coefficients $d_{qm}(t)$. However, we can use the ideas of Richardi and Burrows [30], Wacker [31], and Hansen [10, pp.111-113, pp.140-144] to analytically extend $c_m(\theta, t)$ into a periodic function on the interval $0 \leq \theta \leq 2\pi$. To do this, note that by defining $f_{nm}(\theta) = P_n^m(\cos\theta)$ for $0 \leq \theta \leq \pi$, the analytical extension of $f_{nm}(\theta)$ to the interval $\pi \leq \theta \leq 2\pi$ is given by $(-1)^m f_{nm}(2\pi - \theta)$. [Note that $P_n^m(\cos\theta)$ is not a polynomial of $\cos\theta$ because the expression for $P_n^m(x)$, in general, involves $\sqrt{1-x^2}$. For example, $P_1^1(\cos\theta) = -\sin\theta$.] This shows that the analytical extension of $c_m(\theta, t)$ to the interval $0 \leq \theta \leq 2\pi$ is given by

$$\tilde{c}_m(\theta, t) = \begin{cases} c_m(\theta, t), & 0 \leq \theta \leq \pi \\ (-1)^m c_m(2\pi - \theta, t), & \pi \leq \theta \leq 2\pi \end{cases} . \tag{4.7}$$

Furthermore, $\tilde{c}_m(\theta, t)$ is a periodic spatially bandlimited function on the interval $0 \leq \theta \leq 2\pi$ given by the right side of (4.6) for all θ on the interval $0 \leq \theta \leq 2\pi$. Consequently, the expansion coefficients $d_{qm}(t)$ are given by

$$d_{qm}(t) = \frac{1}{2\pi} \int_0^{2\pi} \tilde{c}_m(\theta, t) e^{-iq\theta} d\theta. \tag{4.8}$$

The sampling theorem for spatially bandlimited periodic functions [10, p.372] and the equation (4.7) then yield the following *exact* expression for the coefficients $d_{qm}(t)$

$$d_{qm}(t) = \frac{1}{2N+1} \left[c_m(0, t) + \sum_{l_\theta=1}^{N} \left\{ e^{-iql_\theta\Delta\theta} + (-1)^m e^{iql_\theta\Delta\theta} \right\} c_m(l_\theta\Delta\theta, t) \right] \tag{4.9}$$

where $\Delta\theta = 2\pi/(2N+1)$, $m = -N, ..., 0, ..., N$, and $q = -N, ..., 0, ..., N$. [Here l_θ is an integer that determines the value of θ at the scan point by the equation $\theta = l_\theta\Delta\theta$.] Inserting the expansion (4.6) for the coefficient $c_m(\theta, t)$ into (4.3) we obtain the Fourier expansion for the time-domain field on the scan sphere:

$$\Phi(a\hat{r}(\theta, \phi), t) = \sum_{m=-N}^{N} \sum_{q=-N}^{N} d_{qm}(t) e^{iq\theta} e^{im\phi}. \tag{4.10}$$

Finally insert (4.10) into the expression (2.7) for the time-domain expansion coefficients $C_{nm}(t)$ to get

$$C_{nm}(t) = \sum_{q=-N}^{N} I_{qnm} d_{qm}(t) \tag{4.11}$$

which expresses the expansion coefficients in terms of sampled near-field data and in terms of the constants

$$I_{qnm} = 2\pi \sqrt{\frac{2n+1}{4\pi} \frac{(n-m)!}{(n+m)!}} \int_0^{\pi} e^{iq\theta} P_n^m(\cos\theta) \sin\theta d\theta \tag{4.12}$$

with $n = 0, 1, ..., n$, $m = -n, ..., 0, ..., n$, and $q = -N, ..., 0, ..., N$. The definition for the associated Legendre functions show that the constants I_{qnm} satisfy the symmetry relations $I_{-q,n,m} = I^*_{qnm}$ and $I_{q,n,-m} = (-1)^m I_{qnm}$.

Having shown how to sample the field on the scan sphere with respect to the spherical angles, we now consider sampling with respect to time. Begin by assuming that the time-domain field on the scan sphere is significantly nonzero only in the time interval $t_0 < t < t_0 + (N_t - 1)\Delta t$, where $\Delta t = \pi/\omega_{max}$ is time-sample spacing, and N_t is an integer determined such that the field on the scan sphere is zero for $t > t_0 + (N_t - 1)\Delta t$. With these assumptions the time-domain field on the scan sphere can be determined from the reconstruction theorem [13, eq.(4.13)], [16], [32, p.83]

$$\Phi(a\hat{r}, t) = \sum_{l_t=0}^{N_t-1} \frac{\sin\left[\pi\left(\frac{t-t_0}{\Delta t} - l_t\right)\right]}{\pi\left(\frac{t-t_0}{\Delta t} - l_t\right)} \Phi(a\hat{r}, t_0 + l_t\Delta t). \tag{4.13}$$

This formula expresses the time-domain field on the scan sphere at all times in terms of its values at the times $t = t_0 + l_t\Delta t$, $l_t = 0, 1, ..., N_t - 1$. It is easily seen that the time-domain coefficients $C_{nm}(t)$ can also be represented by this reconstruction theorem. We have now derived a complete set of formulas that allow us to compute the field outside the scan sphere from space-time samples of the field on the scan sphere. These formulas, which constitute a time-domain sampling theorem, can be written in a very compact form as follows.

First define $\Phi_{l_\theta,l_\phi,l_t}$ to be the acoustic field on the scan sphere at the point $\theta = l_\theta\Delta\theta$, $\phi = l_\phi\Delta\phi$, and $t = t_0 + l_t\Delta t$, that is,

$$\Phi_{l_\theta,l_\phi,l_t} = \Phi(a\hat{r}(l_\theta\Delta\theta, l_\phi\Delta\phi), t_0 + l_t\Delta t). \tag{4.14}$$

The integers l_θ, l_ϕ, and l_t, which determine the space-time sample point, can have the following values: $l_\theta = 0, ..., N$, $l_\phi = 0, ..., 2N$, and $l_t = 0, ..., N_t - 1$. Defining $C_{nml_t} = C_{nm}(t_0 + l_t\Delta t)$ we find from (4.5), (4.9), and (4.11) that

$$\begin{aligned} C_{nml_t} = \tfrac{1}{(2N+1)^2} \sum_{q=-N}^{N} I_{qnm} \sum_{l_\phi=0}^{2N} \Bigg[\Phi_{0,l_\phi,l_t} \\ + \sum_{l_\theta=1}^{N} \left\{e^{-iql_\theta\Delta\theta} + (-1)^m e^{iql_\theta\Delta\theta}\right\} \Phi_{l_\theta,l_\phi,l_t}\Bigg] e^{-iml_\phi\Delta\phi} \end{aligned} \tag{4.15}$$

and (4.13) then shows that the time-domain expansion coefficients are given by

$$C_{nm}(t) = \sum_{l_t=0}^{N_t-1} \frac{\sin\left[\pi\left(\frac{t-t_0}{\Delta t} - l_t\right)\right]}{\pi\left(\frac{t-t_0}{\Delta t} - l_t\right)} C_{nml_t}. \tag{4.16}$$

The formulas (4.15) and (4.16) give the time-domain expansion coefficients $C_{nm}(t)$ for the truncated expansion (4.2) directly in terms of the sampled values (4.14) of the acoustic field on the scan sphere. Thus, to calculate the acoustic field outside the scan sphere one has to know the acoustic field on the scan sphere at $(N+1)(2N+1)N_t$ space-time sample points. The integer N is given by (4.1) and determines the number of spherical modes required for accurate field computation; whereas the integer N_t is determined by the duration of the field on the scan sphere and the time-sample spacing.

The probe-corrected analogs of the formulas derived in this section are obtained by first defining $b_{p,l_\theta,l_\phi,l_t}$ to be the time-domain output of the probe at the point $\theta = l_\theta \Delta\theta$, $\phi = l_\phi \Delta\phi$, and $t = t_0 + l_t \Delta t$. With this definition, the time-domain coefficients $C^p_{nm}(t)$ in (3.11) are given by (4.15) and (4.16) with $\Phi_{l_\theta,l_\phi,l_t}$ replaced by $b_{p,l_\theta,l_\phi,l_t}$.

4.2. Sampling theorems for electromagnetic fields. Assume that the electromagnetic field on the scan sphere is bandlimited so that $\bar{E}_\omega$ can be set equal to zero for $|\omega| > \omega_{max}$. Then the frequency-domain coefficients (2.23) and (2.24) are zero for $|\omega| > \omega_{max}$ and it follows from [10, p.17] that all terms with $n > N$ in the frequency-domain eigenfunction expansions (2.17) are equal to zero. The integer N is the same as in the acoustic case and is given by (4.1).

Taking the inverse Fourier transform of the truncated version of the frequency-domain expansion (2.17) and using the same technique that led to (2.31) it follows that the time-domain electric field for $r \geq a$ is given by the truncated expansion

$$\bar{E}^{tan}(\bar{r},t) = \sum_{n=1}^{N} \Bigg[\frac{a}{r} \sum_{m=-n}^{n} A_{nm}(t - r/c + a/c)\bar{M}_{nm}(\theta,\phi) + B_{nm}(t - r/c + a/c)\bar{N}_{nm}(\theta,\phi)$$

$$- \frac{ic}{a} \sum_{s=1}^{n} \frac{h_n^{(1)}(\omega_{ns} r/c)}{h_n^{(1)\prime}(\omega_{ns} a/c)} \sum_{m=-n}^{n} \int_{t_0}^{t-r/c+a/c} A_{nm}(t') e^{-i\omega_{ns}(t-t')} dt' \bar{M}_{nm}(\theta,\phi)$$

$$- \frac{ic}{a} \sum_{s=1}^{n+1} \frac{g_n(\omega'_{ns} r/c)}{g'_n(\omega'_{ns} a/c)} \sum_{m=-n}^{n} \int_{t_0}^{t-r/c+a/c} B_{nm}(t') e^{-i\omega'_{ns}(t-t')} dt' \bar{N}_{nm}(\theta,\phi) \Bigg] \tag{4.17}$$

where all fields on the scan sphere have been assumed zero for $t < t_0$. [The time-domain far-field pattern is similarly given by (2.32) with the upper limit of the summation over n changed from ∞ to N.]

By a procedure similar to the that used for the acoustic field in Section 4.1, we find the following expressions for the time-domain expansion coefficients derived in [22] and [23]. First define the near-field data

$$
\begin{aligned}
E_{\theta,l_\theta,l_\phi,l_t} &= E_\theta(a\hat{r}(l_\theta\Delta\theta, l_\phi\Delta\phi), t_0 + l_t\Delta t),\\
E_{\phi,l_\theta,l_\phi,l_t} &= E_\phi(a\hat{r}(l_\theta\Delta\theta, l_\phi\Delta\phi), t_0 + l_t\Delta t)
\end{aligned}
\tag{4.18}
$$

and introduce the coefficients

$$
\begin{aligned}
A_{nml_t} = -\tfrac{1}{(2N+1)^2} \sum_{q=-N}^{N} \sum_{l_\phi=0}^{2N} \Bigg\{ im I^A_{qnm} \Bigg[E_{\theta,0,l_\phi,l_t}\\
+ \sum_{l_\theta=1}^{N} \left\{ e^{-iql_\theta\Delta\theta} + (-1)^{m+1} e^{iql_\theta\Delta\theta} \right\} E_{\theta,l_\theta,l_\phi,l_t} \Bigg]
\end{aligned}
$$

$$
\begin{aligned}
&+ I^B_{qnm} \Bigg[E_{\phi,0,l_\phi,l_t}\\
&+ \sum_{l_\theta=1}^{N} \left\{ e^{-iql_\theta\Delta\theta} + (-1)^{m+1} e^{iql_\theta\Delta\theta} biggr\} E_{\phi,l_\theta,l_\phi,l_t} \right] \Bigg\} e^{-iml_\phi\Delta\phi}
\end{aligned}
\tag{4.19}
$$

and

$$
\begin{aligned}
B_{nml_t} = \tfrac{1}{(2N+1)^2} \sum_{q=-N}^{N} \sum_{l_\phi=0}^{2N} \Bigg\{ I^B_{qnm} \Bigg[E_{\theta,0,l_\phi,l_t}\\
+ \sum_{l_\theta=1}^{N} \left\{ e^{-iql_\theta\Delta\theta} + (-1)^{m+1} e^{iql_\theta\Delta\theta} \right\} E_{\theta,l_\theta,l_\phi,l_t} \Bigg]
\end{aligned}
$$

$$
\begin{aligned}
&- im I^A_{qnm} \Bigg[E_{\phi,0,l_\phi,l_t}\\
&+ \sum_{l_\theta=1}^{N} \left\{ e^{-iql_\theta\Delta\theta} + (-1)^{m+1} e^{iql_\theta\Delta\theta} \right\} E_{\phi,l_\theta,l_\phi,l_t} \Bigg] \Bigg\} e^{-iml_\phi\Delta\phi}
\end{aligned}
\tag{4.20}
$$

where the constants I^A_{qnm} and I^B_{qnm} are given by

$$
I^A_{qnm} = 2\pi \sqrt{\frac{2n+1}{4\pi n(n+1)} \frac{(n-m)!}{(n+m)!}} \int_0^\pi e^{iq\theta} P_n^m(\cos\theta)\, d\theta
\tag{4.21}
$$

and

$$
I^B_{qnm} = -2\pi \sqrt{\frac{2n+1}{4\pi n(n+1)} \frac{(n-m)!}{(n+m)!}} \int_0^\pi (iq \sin\theta + \cos\theta) e^{iq\theta} P_n^m(\cos\theta)\, d\theta.
\tag{4.22}
$$

With the help of the reconstruction formulas is has been shown in [22] and [23] that the time-domain expansion coefficients can be written in terms of

A_{nml_t} and B_{nml_t} as

(4.23)
$$A_{nm}(t) = \sum_{l_t=0}^{N_t-1} \frac{\sin\left[\pi\left(\frac{t-t_0}{\Delta t} - l_t\right)\right]}{\pi\left(\frac{t-t_0}{\Delta t} - l_t\right)} A_{nml_t},$$
$$B_{nm}(t) = \sum_{l_t=0}^{N_t-1} \frac{\sin\left[\pi\left(\frac{t-t_0}{\Delta t} - l_t\right)\right]}{\pi\left(\frac{t-t_0}{\Delta t} - l_t\right)} B_{nml_t}.$$

The formulas (4.19)-(4.23) express the time-domain expansion coefficients for the truncated expansion (4.17) directly in terms of the sampled values (4.18) of the tangential electric field on the scan sphere. To calculate the electromagnetic field outside the scan sphere one has to know the tangential electric field on the scan sphere at $(N+1)(2N+1)N_t$ space-time sample points.

To determine the time-domain expansion coefficients $A^p_{nm}(t)$ and $B^p_{nm}(t)$ in (3.32) and (3.33) for the probe-corrected expansion (3.29), first define

(4.24)
$$b_{p\theta,l_\theta,l_\phi,l_t} = b_{p\theta}(a\hat{r}(l_\theta\Delta\theta, l_\phi\Delta\phi), t_0 + l_t\Delta t),$$
$$b_{p\phi,l_\theta,l_\phi,l_t} = b_{p\phi}(a\hat{r}(l_\theta\Delta\theta, l_\phi\Delta\phi), t_0 + l_t\Delta t).$$

Then the desired expansion coefficients $A^p_{nm}(t)$ and $B^p_{nm}(t)$ are given by (4.19)-(4.23) with $E_{\theta,l_\theta,l_\phi,l_t}$ replaced by $b_{p\theta,l_\theta,l_\phi,l_t}$ and $E_{\phi,l_\theta,l_\phi,l_t}$ replaced by $b_{p\phi,l_\theta,l_\phi,l_t}$.

4.3. Numerical example. We shall now present a numerical acoustic example to verify the truncated time-domain expansion (4.2) and the formulas (4.14)-(4.16) for the time-domain expansion coefficients.

Assume that acoustic field is generated by an acoustic point source with Gaussian time dependence located at $\bar{r}_s = (x_s, y_s, z_s)$ whose field is given by

(4.25)
$$\Phi(\bar{r}, t) = \frac{f(t - |\bar{r} - \bar{r}_s|/c)}{4\pi|\bar{r} - \bar{r}_s|}, \quad f(t) = e^{-4t^2/\tau^2}$$

where τ is a positive constant. The spectrum of the Gaussian time function $f(t) = e^{-4t^2/\tau^2}$ is $f_\omega = \tau(4\sqrt{\pi})^{-1}e^{-\omega^2\tau^2/16}$ and in [13, sec.4.3], [16] it was found that the effective maximum frequency for this source is $\omega_{max} = 12/\tau$ and therefore the time-sample spacing is $\Delta t = \pi/\omega_{max} = \pi\tau/12 \approx 0.26\tau$. The minimum wavelength $\lambda_{min} = 2\pi c/\omega_{max} = \pi c\tau/6 \approx 0.5c\tau$ and the maximum wave number $k_{max} = 2\pi/\lambda_{min} = 12/(c\tau)$. It is seen from [13, fig.4.1], [16] that the effective duration of the Gaussian time signal is approximately $T = 2.5\tau$ and that the time signal begins at $t = -1.25\tau$.

In the following we assume that the acoustic point source is located at $\bar{r}_s = 0.1c\tau\hat{x} - 0.4c\tau\hat{y} + 0.2c\tau\hat{z}$ so that the radius of the minimum sphere is

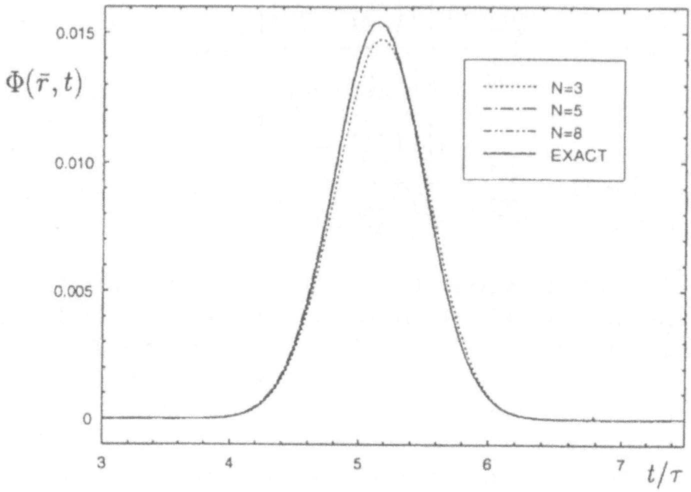

FIG. 4. *The acoustic point-source field at $\bar{r} = 3.0c\tau\hat{x} - 2.1c\tau\hat{y} + 4.1c\tau\hat{z}$ calculated from sampled time-domain spherical near-field data with $N = 3, 5, 8$. Also shown is the exact value of the acoustic point-source field. Its entire time duration is shown.*

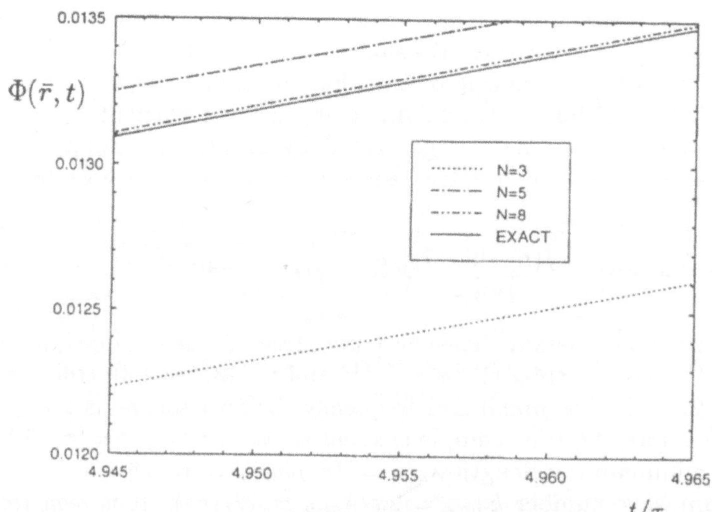

FIG. 5. *The acoustic point-source field at $\bar{r} = 3.0c\tau\hat{x} - 2.1c\tau\hat{y} + 4.1c\tau\hat{z}$ calculated from sampled time-domain spherical near-field data with $N = 3, 5, 8$. Also shown is the exact value of the acoustic point-source field. Only part of its time duration is shown.*

$r_s \approx 0.46c\tau$. With this location of the source the number of spherical modes required for accurate field calculation is found from (4.1) to be $N = 5 + n_1$. In the following we present results for $N = 3, 5, 8$. The scan sphere is chosen to be $r = a = 0.6c\tau$ and consequently the field on the scan sphere is significantly nonzero only when $-1.11\tau < t < 2.31\tau$. Thus, we may chose $t_0 = -1.2\tau$ and $N_t = 15$ so that the field on the scan sphere is sampled at the times $t = -1.2\tau + l_t \cdot 0.26\tau$, for $l_t = 0, 1, 2, ..., 14$.

Figure 4 shows the field at the observation point $\bar{r} = 3.0c\tau\hat{x} - 2.1c\tau\hat{y} + 4.1c\tau\hat{z}$ calculated by (4.2) with $N = 3, 5, 8$ spherical modes. Figure 5 shows the acoustic point-source field in the time interval $4.945\tau < t < 4.965\tau$ and allow us to distinguish the exact result from the result obtained by the spherical near-field formula (4.2) with $N = 8$.

It is seen that the field calculated from the spherical near-field formula indeed approaches the exact field as the number of spherical modes gets larger. For time-domain planar near-field scanning it was found in [13, sec.4.4], [16] that the computed far-field pattern was extended erroneously by an erroneous signal that was caused by the truncated scan plane. Because the spherical scan surface is not truncated no erroneous signal occur in the computed field. Thus, in this respect time-domain spherical near-field scanning is advantageous over time-domain planar near-field scanning.

REFERENCES

[1] A.D. YAGHJIAN, *An overview of near-field antenna measurements*, IEEE Trans. Antenna Propagat., Vol. AP-34, pp. 30-45, Jan. 1986.

[2] J. APPEL-HANSEN, J.D. DYSON, E.S. GILLESPIE, AND T.G. HICKMAN, *Antenna measurements*, Chapter 8 of *The Handbook of Antenna Design*, A.W. RUDGE, K. MILNE, A.D. OLVER, AND P. KNIGHT, Eds., Peter Peregrinus 1982.

[3] D.M. KERNS, *Plane-Wave Scattering-Matrix Theory of Antennas and Antenna-Antenna Interactions*, NBS Monograph 162, 1982.

[4] A.C. NEWELL, *Error analysis techniques for planar near-field measurements*, IEEE Trans. Antenna Propagat., Vol. AP-36, pp. 754-768, Jun. 1988.

[5] W.M. LEACH, AND D.T. PARIS, *Probe compensated near-field measurements on a cylinder*, IEEE Trans. Antenna Propagat., Vol. AP-21, pp. 435-445, Jul. 1973.

[6] A.D. YAGHJIAN, *Near-field antenna measurements on a cylindrical surface: A source scattering-matrix formulation*, NBS Tech. Note 696 (revised), 1977.

[7] F. JENSEN, *Electromagnetic near-field-far-field correlations*, Ph.D. Thesis, Electromagnetics Institute, Technical University of Denmark, Report LD 15, 1970.

[8] P.F. WACKER, *Near-field antenna measurements using a spherical scan: Efficient data reduction with probe correction*, IEE Conf. Publ., Vol. 113, pp. 286-288, 1974.

[9] F.H. LARSEN, *Probe-corrected spherical near-field antenna measurements*, Ph.D. Thesis, Electromagnetics Institute, Technical University of Denmark, Report LD 36, 1980.

[10] J.E. HANSEN, ED., J. HALD, F. JENSEN, AND F.H. LARSEN, *Spherical Near-Field Antenna Measurements*, Peter Peregrinus, 1988.

[11] A.D. YAGHJIAN, *Simplified approach to probe-corrected spherical near-field scanning*, Electronics Letters, Vol. 20, pp. 195-196, Mar. 1984.

[12] A.D. YAGHJIAN AND R.C. WITTMANN, *The receiving antenna as a linear differential operator: Application to spherical near-field measurements*, IEEE Trans.

Antenna Propagat., Vol. AP-33, pp. 1175-1185, Nov. 1985.

[13] T.B. HANSEN AND A.D. YAGHJIAN, *Formulation of time-domain planar near-field measurements without probe correction*, Rome Laboratory In-House Report RL-TR-93-210, 1993.

[14] T.B. HANSEN AND A.D. YAGHJIAN, *Formulation of probe-corrected time-domain planar near-field measurements*, Rome Laboratory In-House Report RL-TR-94-74, 1994.

[15] T.B. HANSEN AND A.D. YAGHJIAN, *Planar near-field scanning in the time domain, Part 1: Formulation*, IEEE Trans. Antenna Propagat., Vol. AP-42, pp. 1280-1291, Sep. 1994.

[16] T.B. HANSEN AND A.D. YAGHJIAN, *Planar near-field scanning in the time domain, Part 2: Sampling theorems and computation schemes*, IEEE Trans. Antenna Propagat., Vol. AP-42, pp. 1292-1300, Sep. 1994.

[17] T.B. HANSEN AND A.D. YAGHJIAN, *Formulation of probe-corrected planar near-field scanning in the time domain*, IEEE Trans. Antenna Propagat., Vol. Ap-43 pp. 569–584, Jan. 1995.

[18] D.S. JONES, *Acoustic and Electromagnetic Waves*, Oxford University Press, 1986.

[19] M. ABRAMOWITZ AND I.A. STEGUN, *Handbook of Mathematical Functions*, Dover, 1972.

[20] J.D. JACKSON, *Classical Electrodynamics*, 2nd edition, Wiley, 1975.

[21] F.W.J. OLVER, *The asymptotic expansion of Bessel functions of large order*, Phil. Trans. Royal Soc. London, Ser. A, Vol. 247, pp. 328-368, 1954

[22] T.B. HANSEN, *Formulation of time-domain spherical near-field measurements*, Rome Laboratory In-House Report, To appear.

[23] T.B. HANSEN, *Formulation of spherical near-field scanning for time-domain electromagnetic fields*, IEEE Trans. Antenna Propagat., In review.

[24] T.B. HANSEN, *Spherical expansions of time-domain acoustic fields: application to near-field scanning*, J. Acoust. Soc. Am., Vol. 98, pp. 1204–1215, Aug. 1995.

[25] J.A. STRATTON, *Electromagnetic Theory*, McGraw-Hill, 1941.

[26] H. ÜBERALL AND G.C. GAUNAURD, *The physical content of the singularity expansion method*, Appl. Phys. Lett., Vol. 39, pp. 362-364, Aug. 1981.

[27] A.J. DEVANEY AND E. WOLF, *Multipole expansions and plane wave representations of the electromagnetic field*, J. Math. Phys., Vol. 15, pp. 234-244, Feb. 1974.

[28] A.D. Yaghjian, *Generalized or adjoint reciprocity relations for electroacoustic transducers*, J. Res. B, Vol. 79B, pp. 17-39, Jan-Jun. 1975.

[29] R.C. WITTMANN, *Probe-corrected spherical near-field scanning theory in acoustics*, IEEE Trans. Instrument. Measurement, Vol. IM-41, pp. 17-21, Feb. 1992.

[30] L.J. RICARDI AND M.L. BURROWS, *A recurrence technique for expanding a function in spherical harmonics*, IEEE Trans. Computer, Vol. C-21, pp. 583-585, Jun. 1972.

[31] P.F. WACKER, *Non-planar near-field measurements: Spherical scanning*, NBS Internal Report 75-809, 1975.

[32] E.O. BRIGHAM, *The Fast Fourier Transform and Its Applications*, Prentice Hall, 1988.

PHASE-SENSITIVE AMPLIFICATION OF PULSES IN NONLINEAR OPTICAL FIBERS

WILLIAM L. KATH*

Abstract. Analysis of the nonlinear Schrödinger (NLS) equation and jump conditions governing pulse propagation in a nonlinear optical fiber with periodically-spaced, phase-sensitive parametric amplifiers (PSAs) shows that the averaged pulse evolution is governed by a fourth-order nonlinear diffusion equation similar to the Kuramoto-Sivashinsky and Swift-Hohenberg equations. A consequence of this diffusive dynamics is that dispersive radiation from evolving pulses is almost totally eliminated, and stable optical pulses propagate over long distances without deformation. Here the asymptotic derivation of this averaged fourth-order equation is reviewed, and comparisons are given between numerical solutions of the averaged equation and the full NLS equation with PSAs. It is shown that the main difference between the two computations is due to a small amount of high-frequency dispersive radiation. The numerical requirements for resolving these high-frequency components is also discussed.

Key words. nonlinear Schrödinger (NLS) equation, solitons, optical fibers, phase-sensitive optical parametric amplifiers, pseudo-spectral methods

AMS(MOS) subject classifications. 35Q51, 35Q55, 65M70, 35C20

1. Introduction. Pulse propagation in a polarization-preserving, single-mode nonlinear optical fiber (see Fig. 1.1) is governed by the one-dimensional nonlinear Schrödinger (NLS) equation [1,2,3,4],

$$i\frac{\partial q}{\partial Z} + \frac{1}{2}\frac{\partial^2 q}{\partial T^2} + |q|^2 q = 0\,. \tag{1.1}$$

Note the standard treatment of pulses in nonlinear optical fibers involves *signaling coordinates*, so that here the dimensionless distance Z and time T are reversed from the standard mathematical convention [2,3,4].

The exact analytical solution of the NLS equation, which is found using the inverse scattering transform (IST) [5,6,7], shows that it possesses nonlinear pulse solutions called *solitons.* Such special pulses pass through one another without permanent change of shape, and are also very robust when disturbed. This resistance to perturbations by solitons is attractive enough that they are being considered for use in high-speed, long-distance optical communication systems [8,9].

Nonlinear optical fibers are particularly convenient to study from the point of view of applied mathematics. First, the NLS equation is derived from Maxwell's equations in a small wavelength (or high frequency) limit.

---

* Department of Engineering Sciences and Applied Mathematics, McCormick School of Engineering and Applied Science, Northwestern University, 2145 Sheridan Road, Evanston, Illinois 60208–3125; Supported in part by the Air Force Office of Scientific Research, 93-1-0084, and by the National Science Foundation, DMS 92-08415 and 93-04397.

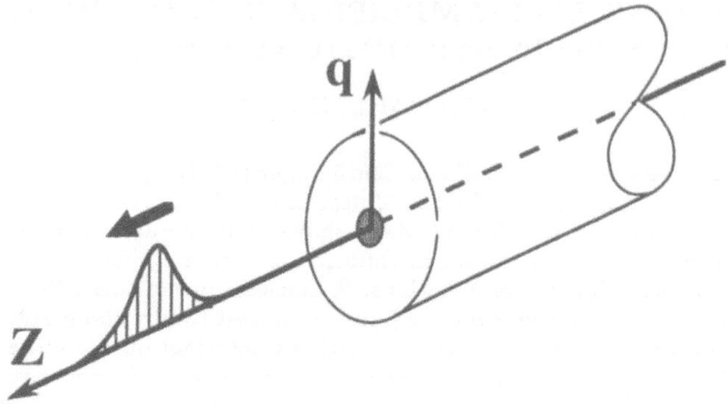

FIG. 1.1. *Schematic of a polarization-preserving, single-mode optical fiber. Note that most of the energy propagating in the Z direction is confined to the inner core of the fiber, which has a slightly larger index of refraction than the surrounding cladding. The amplitude of the electric field envelope is represented by q.*

In realistic situations this limit is approached quite closely, since the wavelength of light is very short in comparison to typical macroscopic dimensions. Thus, the NLS equation usually provides a very good approximation to the actual physical situations it models, a fact that has been verified by a number of experimental results. Second, since the NLS equation is exactly solvable using the inverse scattering transformation, a large amount of detailed information is available as a starting point for the study of applications where perturbed NLS equations are the appropriate models [6,7].

Tremendous strides have been made in reducing loss in optical fibers since their invention, but enough residual loss is present that optical signals must still be amplified every few tens of kilometers. As a result, a fiber/amplifier chain configuration such as that shown in Fig. 1.2 is necessary. For low-dispersion fiber and typical optical pulse widths (e.g., those on the order of 20 ps or greater), however, the natural length scale for the pulse's evolution (either the *dispersion length*, Z_0, or the *soliton period* [2,3]), is several hundred kilometers. This means that the ratio of the amplifier spacing, Z_l, to the dispersion length is a small parameter.

Since the amplifiers' spacing is much smaller than the length scale associated with the pulse's natural evolution, it is possible to average over the shorter of the two [10,11]. The analysis shows that when signal pulses are amplified in a phase-insensitive manner, namely

$$q_{out} = \sqrt{G} q_{in}$$

(where $G \equiv e^{2\alpha}$ is the power gain), and the initial amplitude of the pulses

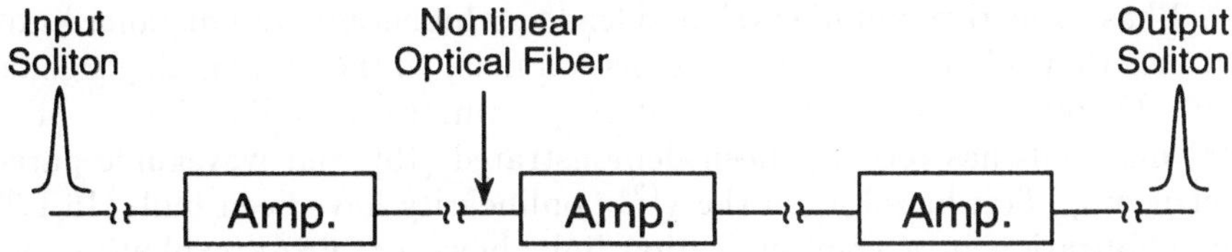

FIG. 1.2. *Schematic of an optical communication line, showing the periodic placement of the optical amplifiers.*

as they are launched into the fiber is larger by the factor

$$\left(\frac{2\alpha}{1-\exp(-2\alpha)}\right)^{1/2}$$

in comparison with the lossless case, then the evolution of the propagating pulses is again governed to leading order by the NLS equation (1.1) [10,11].

Erbium-doped fiber amplifiers are likely to be a common choice for the all-optical amplifiers in the next generation of high-speed systems since they are efficient and relatively inexpensive [12]. These erbium-doped fiber segments amplify signals via stimulated emission; they are pumped with a diode laser to maintain a population inversion, in a manner similar to a laser. The population inversion generates spontaneous emission noise, however, which causes an effect known as Gordon-Haus jitter [13] — the random walk of solitons caused by frequency shifts induced by the noise — and this jitter imposes an overall limit on the maximum allowable bit rate over a given distance, or on the maximum distance signals may propagate at a fixed bit rate.

As a possible alternative to erbium-doped amplifiers, the use of lumped phase-sensitive amplifiers (PSAs) has been suggested [14], which in their simplest implementation amplify pulses according to the rule

$$q_{out} = \mu q_{in} + e^{i\phi}\nu q_{in}^*,$$

where

$$\mu = \frac{1}{2}(\sqrt{G} + \frac{1}{\sqrt{G}}) = \cosh\alpha\,, \qquad \nu = \frac{1}{2}(\sqrt{G} - \frac{1}{\sqrt{G}}) = \sinh\alpha\,,$$

and ϕ is a reference phase associated with the amplifier. Alternatively,

$$q_{out} = e^{i\phi/2}\left(\sqrt{G}q_1 + i\frac{1}{\sqrt{G}}q_2\right),$$

where $q_1 = Re\,(e^{-i\phi/2}q_{in})$ and $q_2 = Im\,(e^{-i\phi/2}q_{in})$. Note that the maximum gain is obtained for the part of the signal that is in phase with the amplifier ($arg\,q_{in} = \phi/2$), while the part of the signal that is out of phase is attenuated.

Phase-sensitive amplifiers should lead to a higher bit-rate-distance limit because they add no spontaneous emission noise to the propagating signals [14]. Quantum-limited phase-sensitive parametric amplification in bulk $\chi^{(2)}$ materials has recently been demonstrated [15], and waveguide parametric amplifiers based upon the $\chi^{(2)}$ nonlinearity have been built [16,17]. For application in a communication link, however, PSAs exploiting the nonlinear refractive index n_2 of the fiber potentially hold more promise [18,19].

We have recently completed the first theoretical studies of pulse evolution in a nonlinear optical fiber where linear loss is balanced by a chain of periodically-spaced PSAs [20,21]. The results show that pulse propagation is stabilized by the use of these amplifiers. This stabilization results because the PSAs eliminate phase variations across a pulse's profile, thus strongly attenuating the effects of dispersion and self-phase modulation. Since this elimination of phase variations does not depend upon nonlinear self-phase modulation being present, we have also been able to show that some of these stabilizing effects are present even for linear pulses. This has led us to suggest the use of PSAs as a new method for compensating dispersion in linear optical fiber systems [22,23]. More recently, we have completed preliminary studies involving the application of PSAs in optical-memory loops [24], and how they can be used to compensate the Raman scattering-induced soliton self-frequency shift [25].

2. Pulse evolution using phase-sensitive amplifiers. Pulse propagation through an optical fiber including periodic parametric amplification, nonlinearity, and linear loss is governed by the equation [20,21]

$$\frac{\partial q}{\partial Z} = \frac{i}{2}\frac{\partial^2 q}{\partial T^2} + i|q|^2 q + \frac{1}{\epsilon} h\left(\frac{Z}{\epsilon}\right) q + \frac{1}{\epsilon} f\left(\frac{Z}{\epsilon}\right) q^*, \tag{2.1}$$

where the rapidly varying functions h and f, namely

$$h(\zeta)q = -\Gamma q + (\cosh\alpha - 1)\sum_{n=1}^{N}\delta(\zeta - nl)q(nl^-), \tag{2.2a}$$

$$f(\zeta)q^* = e^{i\phi}\sinh\alpha\sum_{n=1}^{N}\delta(\zeta - nl)q^*(nl^-), \tag{2.2b}$$

($\zeta = Z/\epsilon$) account for both the uniform loss of the fiber and the effect of the optical phase-sensitive gain of the lumped parametric amplifiers [20,21]. The length Z in Eq. (2.1) has been scaled on a typical dimensional dispersion length Z_0 (e.g., 250–500 km); the dimensional amplifier spacing Z_l is assumed to be much shorter than this length (e.g., $Z_l = 20$–50 km). Mathematically, this assumption is made by defining $\epsilon l = Z_l/Z_0$ and taking $\epsilon \ll 1$, $l \sim \mathcal{O}(1)$ [20,21].

Leading-order averaging shows that, as expected, for the gain at an amplifier to compensate the loss encountered as the signal propagates from one amplifier to the next, one must set $\alpha = \Gamma l$. When a higher-order averaging is performed on Eq. (2.1) with a multiple scale expansion [26] using the short length scale $\zeta = Z/\epsilon$, the dispersion length scale Z, and the long length scale $\xi = \epsilon l Z/2 \tanh \Gamma l$, the following averaged fourth-order evolution equation is obtained [20,21]:

$$\begin{aligned}\frac{\partial U}{\partial \xi} + \frac{1}{4}\frac{\partial^4 U}{\partial T^4} - \frac{\kappa}{2}\frac{\partial^2 U}{\partial T^2} + \left(\frac{\kappa^2}{4} - \Delta\alpha\right) U \\ - \kappa U^3 + U^5 + \beta_1 U \left(\frac{\partial U}{\partial T}\right)^2 + \beta_2 U^2 \frac{\partial^2 U}{\partial T^2} = 0 \,. \end{aligned} \tag{2.3}$$

Here $\beta_1 = 6 - 3\tanh(\Gamma l)/\Gamma l$, $\beta_2 = 3 - \tanh(\Gamma l)/\Gamma l$, $\Delta\alpha$ is an $\mathcal{O}(\epsilon^2)$ deviation from the exact balance between amplification and decay [the full expression for the gain at an amplifier is $\alpha = \Gamma l + (Z_l/Z_0)^2 (\Delta\alpha)/(2\tanh \Gamma l)$], $\kappa = d\phi/dZ$, and $U = [(1-\exp(-2\Gamma l))/(2\Gamma l)]^{1/2} Re\,(q \exp(-i\phi/2))$ is the scaled amplitude of the in-phase quadrature after each amplifier. This fourth-order diffusion equation is reminiscent of the Kuramoto-Sivashinsky and Swift-Hohenberg equations [27]. Note that the parameter κ, which represents a constant amplifier phase rotation rate, can be taken to be unity without loss of generality since it can be scaled out of (2.3).

An idea of the general structure of the solutions to (2.3) can be obtained by considering the limit $\Gamma l \to 0$. In this limit, $U = \eta \operatorname{sech} \eta T$, where $\eta = (1 \pm 2(\Delta\alpha)^{1/2})^{1/2}$, and these solutions are plotted in Fig. 2.1. The limit of small Γl corresponds physically to assuming that the amplifier spacing is small in comparison with the characteristic length scale associated with the linear loss. This particular limit is somewhat unphysical for the application considered here, but it is relevant when shorter pulses and higher bit-rates are considered [25].

For small Γl one can determine the solution structure analytically using perturbation theory. The stability of the various solutions can also be obtained by this method. Of particular interest are the regions near the limit and bifurcation points (regions A and B in Fig. 2.1). The same basic solution structure is found to persist for a wide range of values of Γl [28] (note that the limit point occurs at $\Delta\alpha > 0$ when $\Gamma l > 0$, however).

The perturbation analysis is possible because the basic equation one obtains when (2.3) is linearized about $U = \eta \operatorname{sech} \eta T$ is

$$\tilde{U}_\xi + L_- L_+ \tilde{U} = 0\,, \tag{2.4}$$

where the operators L_- and L_+ are the real and imaginary parts of the linearized NLS equation [29],

$$L_+ = -\frac{1}{2}\partial_T^2 - 3U_0^2 + \frac{1}{2}, \qquad L_- = -\frac{1}{2}\partial_T^2 - U_0^2 + \frac{1}{2}. \tag{2.5}$$

These operators are well understood [29]. Since L_- and L_+ do not commute, however, the linearized operator $-L_- L_+$ is non-selfadjoint and less

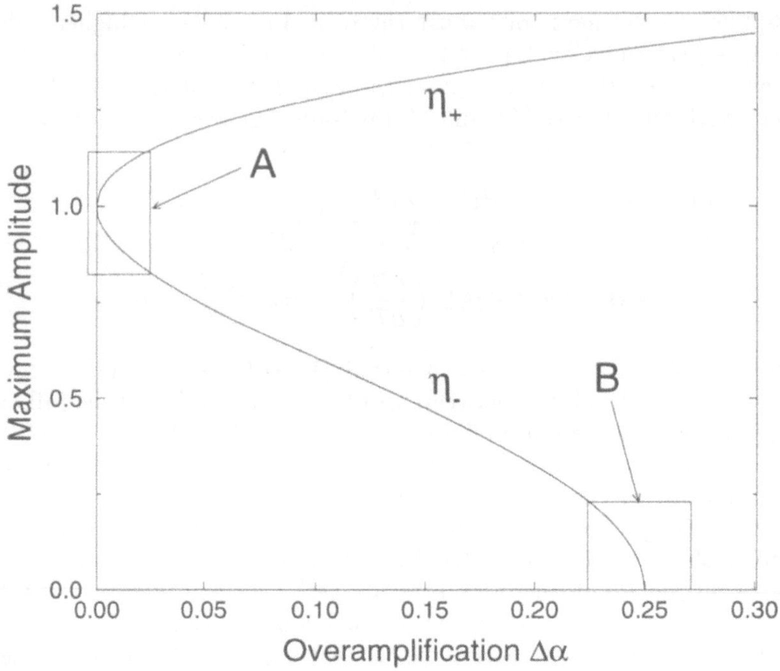

FIG. 2.1. *Plot of the maximum amplitude for the two solution branches* $U = \eta \operatorname{sech} \eta T$ *where* $\eta = (1 \pm 2(\Delta\alpha)^{1/2})^{1/2}$. *Here* $\Gamma l = 0$ *and* $\kappa = 1$.

straightforward to analyze. Using the properties of L_- and L_+ [29], however, it can be shown that $-L_- L_+$ possesses two zero eigenfunctions. Furthermore, it is possible to show that, except for initial conditions which lie in the space spanned by the two zero eigenfunctions, all solutions of the homogeneous part of the linearized equation (2.4) decay exponentially [28].

To determine the stability of (2.3), therefore, it is only necessary to determine the effect of the perturbation upon the two zero eigenvalues. One of these eigenvalues arises due to the translation invariance of (2.3). Since the equation is translation invariant for all values of Γl, this eigenvalue is unaffected by the perturbation. The other zero eigenvalue is affected by the perturbation, however, since this zero eigenmode arises at $\Gamma l = 0$ from an infinitesimal amplitude invariance due to the vertical tangent at the limit point, Region A in Fig. 2.1. Thus, the stability of the pulse is determined solely by whether the perturbation makes this particular eigenvalue positive or negative.

The perturbation analysis of this eigenvalue is relatively straightforward [28], and one finds that when Γl is small, and

$$\frac{4}{405}(\Gamma l)^4 \equiv \Delta\alpha_c < \Delta\alpha < \frac{\kappa^2}{4},$$

a stable steady-state pulse solution of (2.3) exists (e.g., the upper branch in Fig. 2.1). The critical value $\Delta\alpha_c$ determines the minimum amount of overamplification necessary for stable pulse solutions to exist. The need for a small amount of overamplification is consistent with the use of PSAs since there is a small amount of energy lost by the pulse due to the attenuation of the out-of-phase quadrature. For values of $\Delta\alpha$ below $\Delta\alpha_c$, pulses decay to zero. Of course, these analytical results are only valid when Γl is small, but they are nonetheless indicative of the results obtained using numerical simulations for values of Γl which are $O(1)$ [20,21,28].

3. Numerical solutions. When Γl is $\mathcal{O}(1)$ the preceding perturbation analysis is not quantitatively valid, but the evolution and stability of a propagating pulse can be determined numerically. The numerical scheme employed to solve (2.1) or (2.3) utilizes a fourth order Runge-Kutta integration in Z or ξ, and a filtered pseudo-spectral (Fourier) method in T [30,31]. This procedure combines many of the advantages of split-step [3,32,33] and explicit Runge-Kutta [34] methods, giving a relatively simple fourth-order scheme with improved numerical stability properties. In particular, the inherent stiffness of (2.3) at high frequencies is eliminated using this method, allowing much larger integration steps in ξ to be taken than would otherwise be possible [31]. This reduces the required computational resources significantly, which is important since pulse stability over long distances must be examined.

In all of the numerical runs the computational region was taken to be larger than the region of interest, and an absorbing boundary layer was added to eliminate any reflections from the edges of the computational region [35]. The results were carefully checked by varying the number of Fourier modes, the integration step size, and the size of the computational region.

Fig. 3.1 shows two representative numerical solutions of Eq. (2.3) [21]. Fig. 3.1a is for an initial pulse $U(T,0) = \operatorname{sech} T$ and Fig. 3.1b for $U(T,0) = 1.8 \operatorname{sech} T$. In both cases the solution exponentially approaches a stable steady state as it evolves. The parameters used in this computation are $Z_0 = 500$ km, $\Gamma l = 1$ (which corresponds to an amplifier spacing of roughly 36 km for a power loss rate of 0.24 dB/km), $\kappa = 1$, and $\Delta\alpha = 0.1$. As expected from the stability analysis discussed in the previous section, positive values of $\Delta\alpha$ (i.e., overamplification) are necessary to obtain the stable pulse solutions. The pulses in these simulations propagate 10 units in the long length scale ξ, which corresponds physically to a pulse traveling through 2,906 amplifiers for a total distance of roughly 210 dispersion lengths. As mentioned above, the long distance was chosen to explicitly show the stability of the pulses.

A measure of the accuracy of the averaged evolution equation, (2.3), is obtained by comparing its solutions with numerical solutions of the full nonlinear Schrödinger equation with loss and periodic phase-sensitive am-

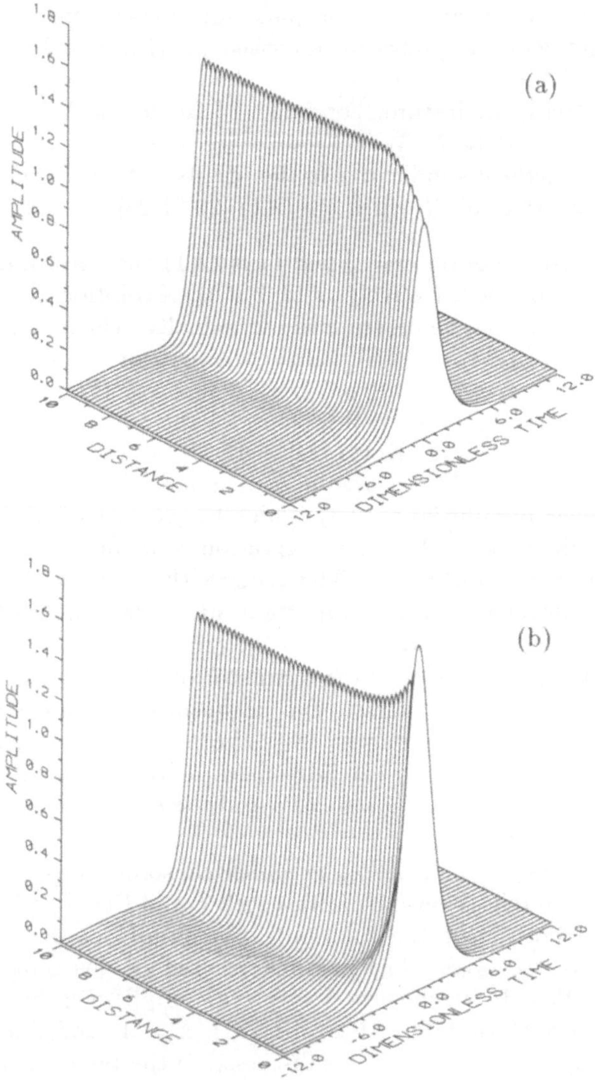

FIG. 3.1. *Evolution of initial pulses* $U(T,0) = \operatorname{sech} T$, *(a), and* $U(T,0) = 1.8 \operatorname{sech} T$, *(b), showing exponential decay onto the stable pulse solution. The parameters are:* $\Gamma l = 1$ *(corresponding to an amplifier spacing of 36 km),* $\kappa = 1$, *and* $\Delta\alpha = 0.1$. *The computations were run to* $\xi = 10$ *(i.e., a distance of 105,000 km for a 500 km dispersion length, or 26,250 km for a 250 km dispersion length) to explicitly show the stability of the pulses.*

plification, (2.1) and (2.2). Fig. 3.2a shows such a comparison for a total propagation distance of 10,000 km, using the same initial pulse and physical parameters as those used for the simulation shown in Fig. 3.1a [21]. Because the two solutions are indistinguishable when plotted together, the difference between the two pulses is shown in Fig. 3.2b. Note that the difference is quite small, of the order 10^{-4}, demonstrating that the averaged equation is indeed an accurate approximation.

The majority of the difference between the two solutions can be attributed to second-order terms which have been ignored in the derivation of (2.3) [21]. In addition, a small amount of linear dispersive radiation can be seen in Fig. 3.2b. This radiation appears to be largest in the vicinity of the main pulse and decreases away from it. This linear dispersive radiation does not show up in the multiple-scale expansion used to derive (2.3) because it is believed to be exponentially small in the perturbation parameter [36]; such exponentially small terms typically do not appear in perturbation expansions using powers of the small parameter [26] unless special techniques are employed [37]. This dispersive radiation will be discussed in more detail in the next section.

It is also illustrative to directly examine the stabilizing effect of the amplifiers by plotting the magnitude of the out-of-phase quadrature between the amplifiers, as determined by numerically solving (2.1) and (2.2). This is shown in Fig. 3.3, which provides clear evidence that after an amplifier the out-of-phase quadrature grows due to forcing from the dispersion and nonlinear self-phase modulation, but that upon reaching the next amplifier it is sharply attenuated. (Note that in this figure the exponential decay due to loss between the amplifiers has been factored out.)

The amplifiers' suppression of the effects of dispersion and self-phase modulation means that the change in the pulse's shape from one amplifier to the next is reduced, and as a result, much longer distances are necessary for a true steady state to be reached. In the numerical simulations shown above in Figs. 3.2 and 3.3, for the example, the solution is not yet close to a steady state, but still in a transient regime.

An example showing the full approach to the steady state is given in Fig. 3.4, where the value at the center of the pulse just after an amplifier is plotted as a function of distance (in dispersion lengths). Results from both the averaged equation, (2.3), and the full NLS simulations, Eqs. (2.1) and (2.2), are shown. The curves are again almost indistinguishable. Note that the solution is not even close to the steady state until the pulse has propagated approximately 100 dispersion lengths, showing the degree to which the phase-sensitive amplifiers are able to eliminate the effects of dispersion and self-phase modulation.

4. Dispersive sideband radiation. Not all frequencies are present in the linear dispersive radiation seen in Fig. 3.2b. Only certain frequencies, or sidebands, are able to maintain phase-matching with the amplifiers as

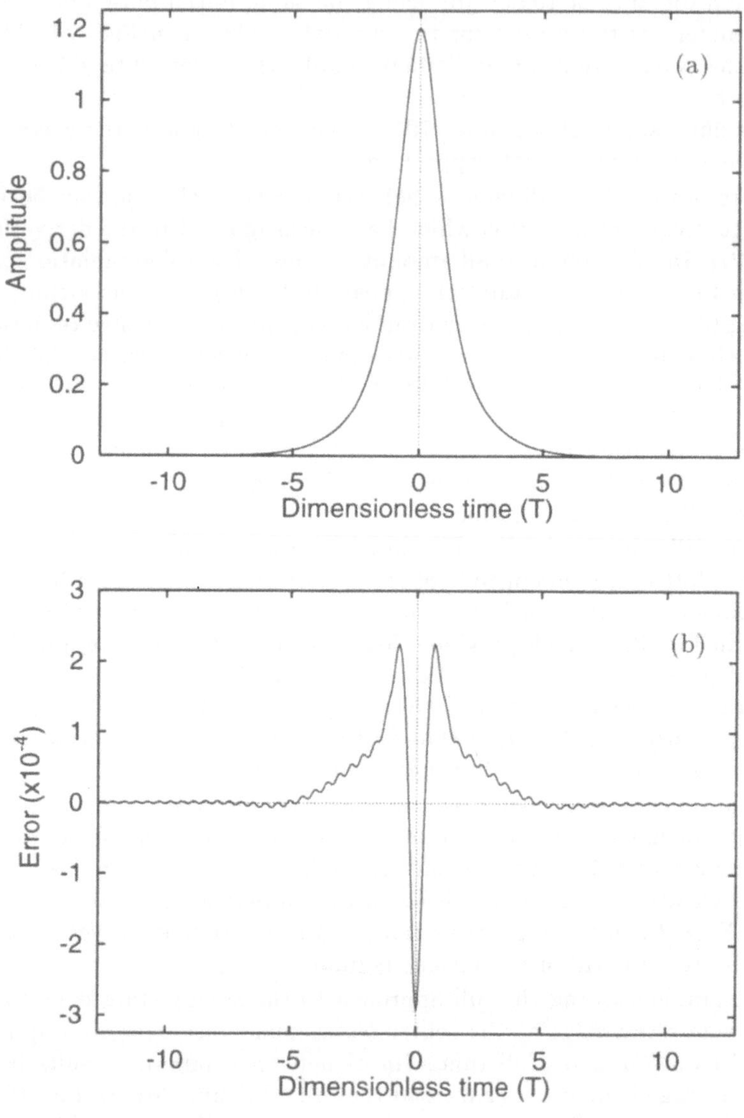

FIG. 3.2. *Comparison of the solutions of the averaged evolution equation and the full nonlinear Schrödinger equation with loss and periodic phase-sensitive amplification, showing the in-phase quadrature, (a), and the difference, (b), between the two solutions. The parameters are $\Gamma l = 1.0$, corresponding to an amplifier spacing of 36 km, a dispersion length of 500 km, $\kappa = 1$, and $\Delta\alpha = 0.1$. The solutions are plotted after a total propagation distance of 10,000 km or 275 amplifiers.*

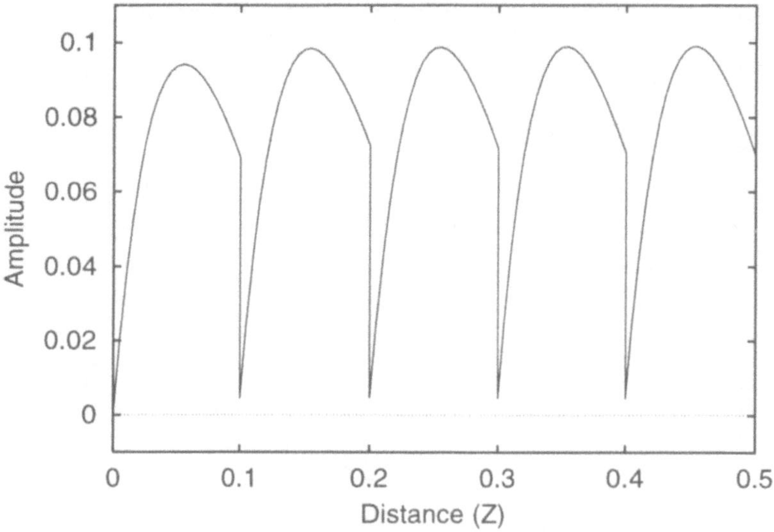

FIG. 3.3. *Midpoint value of the out-of-phase quadrature plotted as a function of distance, showing the evolution between the amplifiers (calculated using the full nonlinear Schrödinger equation with loss and periodic phase-sensitive amplification). The exponential decay due to loss between the amplifiers has been factored out. The magnitude grows after an amplifier, but upon reaching the next one it is sharply attenuated. Here the amplifier spacing is 50 km, and the distance is in terms of dispersion lengths ($Z = 1$ corresponds to 500 km).*

they propagate, and thus only these sidebands experience an overall gain close to unity as they pass through an optical fiber/PSA segment. This can be seen most easily by neglecting the nonlinear term in (2.1). In this case, Fourier components of the form $e^{\pm i\omega T}$ at the beginning of one fiber segment will be transformed into

$$e^{-\Gamma l} e^{-i\omega^2 Z/2 \pm i\omega T}$$

at the end of that segment, just before the next amplifier. These Fourier components will phase-match with the amplifier if the phase advance induced in traveling from one amplifier to the next is the same as the phase advance of the amplifiers, namely

$$-\frac{1}{2}\omega^2 \epsilon l = \frac{1}{2}\epsilon\kappa l \quad (\mathrm{mod}\ 2\pi)\,,$$

or, equivalently,

$$\omega_m^2 = \frac{4m\pi}{\epsilon l} - \kappa\,. \tag{4.1}$$

Note that the $m = 0$ case is not correct since in this particular instance the phase advance from the nonlinear term is significant and should not

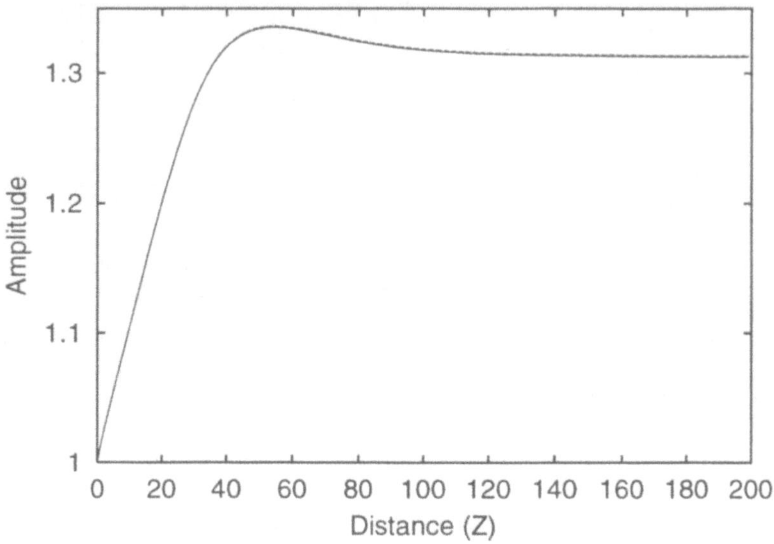

FIG. 3.4. *Midpoint value of the in-phase quadrature just after an amplifier plotted as a function of distance (in terms of dispersion lengths). Results from both the averaged equation (dashed curve), Eq. (2.3), and the full NLS equation with loss and periodic phase-sensitive amplification (solid curve) are plotted. The two are virtually indistinguishable. The parameters are $\Gamma l = 1.0$, corresponding to an amplifier spacing of 36 km, a dispersion length of 500 km, $\kappa = 1$, and $\Delta\alpha = 0.1$. A total propagation distance of 2750 amplifiers or 200 dispersion lengths is shown. Note that an approximate steady-state is not reached until after the pulse has propagated roughly 100 dispersion lengths, showing the degree to which PSAs are able to suppress the effects of dispersion and self-phase modulation.*

be neglected (specifically, the nonlinear term provides an additional phase advance of size κ, which then gives $\omega_0 = 0$). For $m = 1, 2, \ldots$, however, this accurately predicts the frequencies which are present in the numerical solution of (2.1).

For small ϵ the sidebands given by (4.1) occur at high frequency, which explains why the averaging method used to derive (2.3) does not capture them: the averaging method implicitly assumes that all T derivatives remain $\mathcal{O}(1)$ in the limit $\epsilon \to 0$.

It should be noted that the frequencies at which amplification occurs are strongly dependent upon the spacing between the amplifiers. In a realistic situation, therefore, where the amplifier spacing varies somewhat with distance along the fiber line, or the PSA bandwidth is limited, this linear dispersive radiation is expected to be automatically eliminated [38]. For the calculations presented here the amplifier spacing was taken to be exactly periodic and the PSA bandwidth was assumed infinite, and thus the dispersive radiation can survive. The effect of variable amplifier spacing upon the solitons is expected to be minimal, however, because the nonlinear phase shift produced by the solitons is sufficient to keep them phase-locked to the the PSAs.

It is possible to give a more detailed approximation describing the sideband evolution, at least in the linear case. One first defines

$$\hat{u}_n(\omega) = e^{-in\kappa z_l/2}\hat{u}(nz_l^+, \omega)\,,$$

where $\hat{u}(Z,\omega)$ is the Fourier transform of $u(Z,T)$ and $z_l = \epsilon l$, and defines

$$x_n = \left[\hat{u}_n(\omega) + \hat{u}_n^*(-\omega)\right]/2\,, \tag{4.2}$$

$$y_n = \left[\hat{u}_n(\omega) - \hat{u}_n^*(-\omega)\right]/2\,. \tag{4.3}$$

Note that x_n and y_n are then merely the Fourier transforms of the real and imaginary parts of $e^{-in\kappa z_l/2}u(nz_l, T)$ just after an amplifier. Then the linear evolution of the field from amplifier to amplifier, including dispersion and loss between the amplifiers, reduces to the coupled difference equations

$$\begin{pmatrix} x_n \\ y_n \end{pmatrix} = e^{-\Gamma l}\begin{bmatrix} e^{\alpha}\cos\frac{\vartheta}{2} & -ie^{\alpha}\sin\frac{\vartheta}{2} \\ ie^{-\alpha}\sin\frac{\vartheta}{2} & e^{-\alpha}\cos\frac{\vartheta}{2} \end{bmatrix}\begin{pmatrix} x_{n-1} \\ y_{n-1} \end{pmatrix}, \tag{4.4}$$

where α is the amplifier gain and $\vartheta = \kappa z_l + \omega^2 z_l$ [23].

The long-distance pulse evolution in the fiber/PSA line is determined by the eigenvalues of the 2×2 matrix in the above equation, which are

$$\lambda = e^{-\Gamma l}\left[\cos\frac{\vartheta}{2}\cosh\alpha \pm \sqrt{\cos^2\frac{\vartheta}{2}\cosh^2\alpha - 1}\,\right]. \tag{4.5}$$

When $\cos\vartheta = 1$, one of these eigenvalues is $e^{-\Gamma l+\alpha}$, and the other is $e^{-\Gamma l-\alpha}$. This is another way of showing that the loss in one of the two quadratures is compensated by the PSAs (the x quadrature, with exact compensation when $\Gamma l = \alpha$), while the other quadrature (y) is attenuated by *both* the loss in the fiber and the PSAs. After many amplifiers have been passed, the only components of the pulse that survive are those which correspond to an eigenvalue in the above difference equation that is close to 1. Thus, it makes sense to look at frequency ranges which make

$$\frac{\vartheta}{2} = \frac{1}{2}\kappa z_l + \frac{1}{2}\omega^2 z_l \approx 2m\pi\,.$$

Note that these give exactly the same phase-matching frequencies as (4.1).

For simplicity, only one sideband ($m = 1$) will be considered. The frequency ω is assumed to be close to the sideband value, namely $\omega = \omega_1 + \Omega$, so that one can define

$$x_n(\omega) = X_n(\omega - \omega_1) = X_n(\Omega).$$

Since y_n will be small, one has $\hat{u}_n(\omega) \approx \hat{u}_n^*(-\omega)$. This means that

$$u_n(T) \approx \cos(\omega_1 T)\ \frac{1}{\pi}\int_{-\infty}^{\infty} X_n(\Omega)e^{-i\Omega T}\, d\Omega\,, \tag{4.6}$$

and $X_n(\Omega)$ represents the Fourier transform of the sideband envelope. In addition, α is assumed to be close to Γl, i.e., $\alpha = \Gamma l + z_l\alpha_1 + \dots$. When the eigenvalue closest to 1 is expanded near the first sideband frequency, one then obtains

$$\lambda \approx 1 + z_l\alpha_1 - \frac{1}{2}\, z_l^2\omega_1^2\Omega^2 \coth\Gamma l + \dots .$$

Since $X_n = \lambda^n X_0$, this means that

$$X_n \approx \left[1 + \alpha_1 z_l - \frac{1}{2}\, z_l^2\omega_1^2\Omega^2 \coth\Gamma l + \dots\right]^n X_0$$

or, upon using $(1+x)^n \approx e^{xn}$ for small x and large n, $Z = nz_l$, and $\omega_1^2 z_l = 4\pi - \kappa z_l \approx 4\pi$,

$$X(Z,\Omega) \approx e^{\alpha_1 Z - D\Omega^2 Z}\ X(0,\Omega) \tag{4.7}$$

where $D = 2\pi\coth\Gamma l$. This is the solution (in the frequency domain) of the diffusion equation

$$\frac{\partial R}{\partial Z} = \alpha_1 R + D\frac{\partial^2 R}{\partial T^2}\,. \tag{4.8}$$

This equation and its solution (4.7) give a good approximation for the evolution of the sidebands (with suitable modification, of course, for the sidebands with $m > 1$). An example comparison between the approximate, (4.7), and numerical solutions of the linear evolution for a particular sideband is shown in Fig. 4.1. Note that the sidebands do not evolve slowly with distance, as the main pulse does, but rather with an $\mathcal{O}(1)$ change in the envelope occurring over an $\mathcal{O}(1)$ distance in Z.

It is important to make sure that these high-frequency sidebands are being computed correctly, and resolving them properly imposes a number of restrictions on a straightforward pseudo-spectral numerical solution of (2.1) and (2.2). First, since the sidebands are high-frequency components, resolving them properly in T means that a large number of Fourier modes must be used. Second, after many amplifiers the overall gain drops off rapidly as one moves away from one of these sideband frequencies (the overall gain curve becomes quite narrow around each of them), which means that sufficient resolution in frequency (because $\Delta\omega = 2\pi/T_w$) necessitates a large time window, T_w. (It also helps to tune the time window so that the first sideband frequency lies precisely at the frequency of one of the Fourier modes so that the proper maximum gain is captured.) Finally,

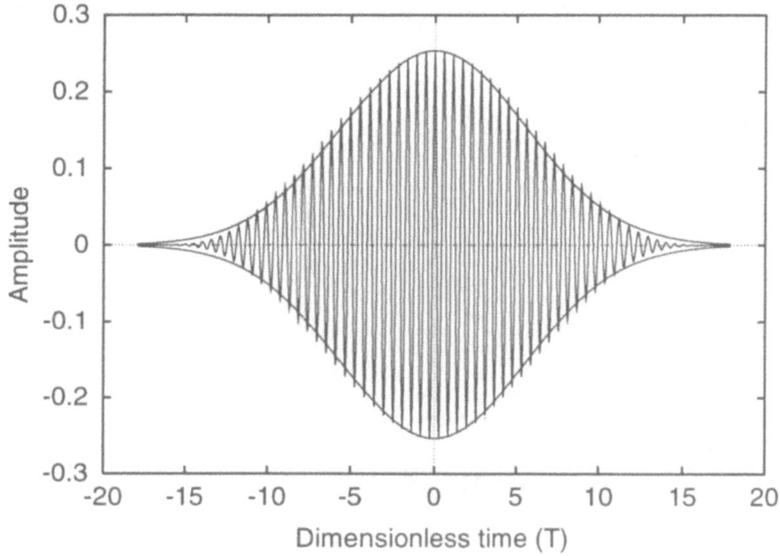

FIG. 4.1. *Comparison between the numerical solution for the linear evolution of a single high frequency sideband in the presence of PSAs, and the approximate envelope solution, (4.7). The sideband frequency is* $\omega_1 = 11.21$, $z_l = 0.1$, $\Gamma l = 0.46052$, $\alpha = 0.463$, *and the solution is plotted at* $Z = 1$. *The initial condition used was* $\cos(\omega_1 T)\exp(-T^2/4)$, *and it is seen that the approximate envelope agrees quite well with the numerical solution.*

since the high frequency components no longer evolve slowly with distance, so that $O(1)$ changes in the solution occur over an $O(1)$ distance in Z, this means that any errors produced by the absorbing layers near the boundary of the computational domain can propagate much more quickly into the interior and affect the dynamics of the rest of the solution. Increasing the width or height of the absorbing layer does not help, however, since if too much additional damping is included, the dynamics of the sidebands can be artificially affected.

As an example, in Fig. 4.2 the results of a highly resolved computation are shown. In this case the ratio of the amplifier spacing to the dispersion length has been taken to be larger, so that the pulse is driven more strongly by the amplifiers. The shedding of dispersive radiation from the pulse is clearly seen. In addition, careful examination of the solution shows that the radiation decreases away from the pulse merely because it grows first near the pulse and spreads slowly with distance; the dispersive radiation farther away from the pulse was generated at an earlier point in the computation, when it had a smaller amplitude.

Fig. 4.3 gives the solution spectrum associated with the pulse shown in Fig. 4.2, and clearly shows the sidebands which have been amplified by the periodically-spaced, phase-sensitive amplifiers. The numerically computed sideband frequencies are very close to the values predicted by (4.1). Fig. 4.3

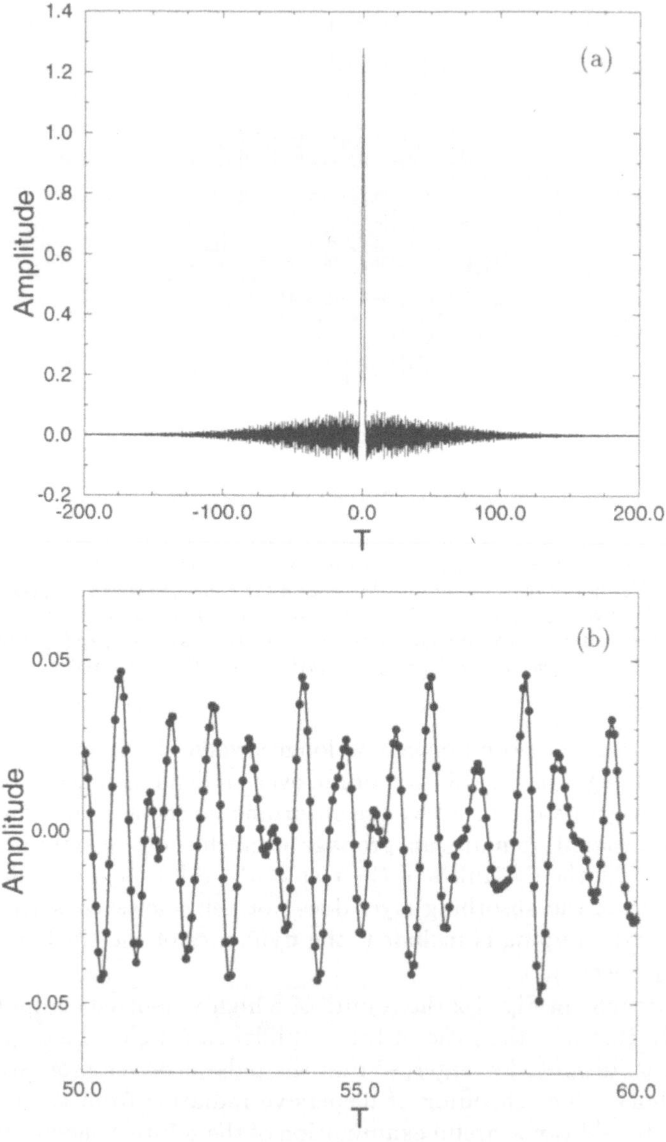

FIG. 4.2. *Plot of the in-phase quadrature for a highly-resolved numerical solution of the full NLS equation with phase-sensitive amplifiers, showing the development of the high-frequency dispersive radiation. The full pulse is shown in (a), and a blown-up section of the dispersive radiation is given in (b) to show that the radiation is adequately resolved in the time-domain ($\Delta x \approx 0.05$). For these computations the parameters are $\Gamma l = 1.105$, corresponding to an amplifier spacing of 40 km, the dispersion length was 200 km, $\kappa = 1$, and $\Delta\alpha = 0.1$. The solutions are plotted after a total propagation distance of 200 dispersion lengths, or 1,000 amplifiers.*

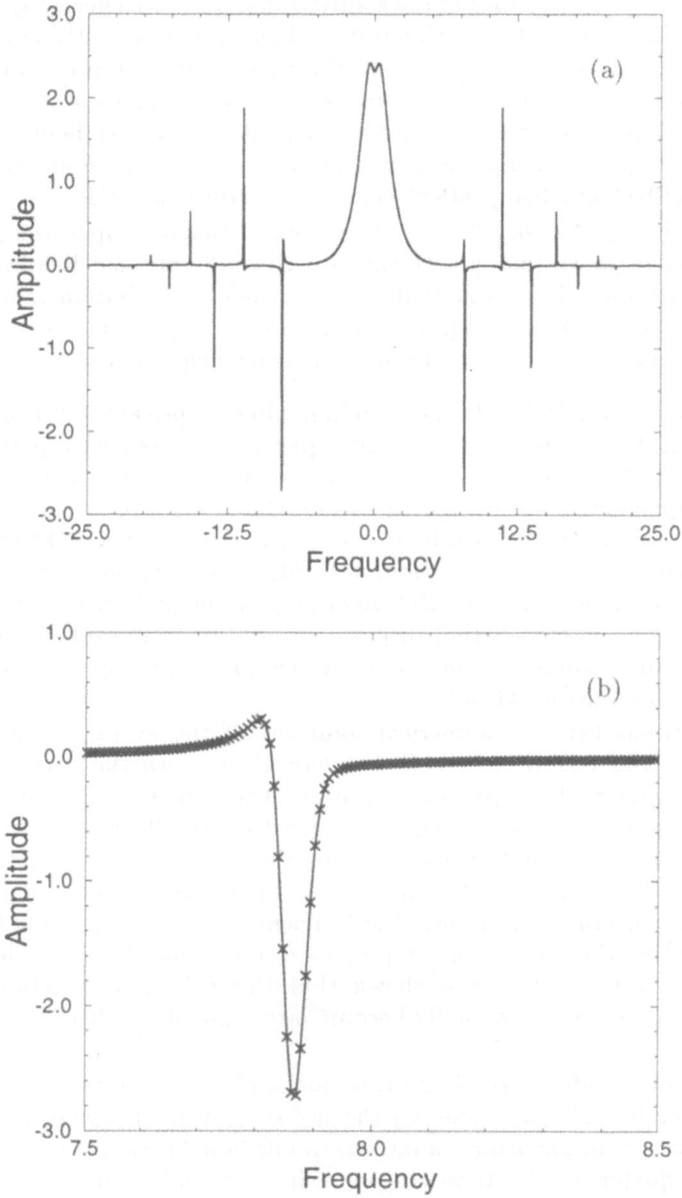

FIG. 4.3. *Plot of the spectrum associated with the solution shown in Fig. 4.2. The full spectrum is shown in (a), with the dispersive sidebands clearly visible. The resolution of the first sideband at* $\omega \approx 7.86$ *is shown in (b). In the computations a spatial width of 818.23 was employed to resolve the solution properly in the frequency domain (note this gives* $\Delta\omega = 0.00768$*, so that the first sideband occurs at mode number 1024), and 16,384 pseudo-spectral Fourier modes were used.*

also shows that large portions of the solution spectrum are effectively zero. This means that much of the computational effort is apparently unnecessary, although it is not possible to skip the computation of any frequency components when a straightforward pseudo-spectral implementation with FFTs and equally-spaced frequencies is used. It is currently being investigated, however, whether or not it is possible to use fractional Fourier transforms (FRFTs) [39,40] rather than more traditional FFTs.

Finally, in Fig. 4.4 the effect of the growing sidebands upon the propagating pulse is shown (along with the growth of the first sideband in the frequency domain). It is seen that as the dispersive radiation grows, it extracts energy from the main pulse, which causes the pulse to decay. This behavior would not be expected from the averaged equation alone.

5. Discussion. Pulse propagation in nonlinear optical fibers has been considered in the case where periodically-spaced, phase-sensitive parametric amplifiers (PSAs) are used to compensate the linear loss in the fiber. Analysis employing multiple-scale expansions shows that the averaged pulse dynamics is no longer governed by the nonlinear Schrödinger (NLS) equation, but rather by a fourth-order nonlinear diffusion equation similar to the Kuramoto-Sivashinsky or Swift-Hohenberg equations [27]. A consequence of this diffusive dynamics is that dispersive radiation from evolving pulses is almost totally eliminated, and stable optical pulses propagate over long distances without deformation.

Comparisons between numerical solutions of the averaged equation and the full NLS equation with PSAs were given, with the two in very good agreement over large parameter ranges. The main difference between the two computations was shown to be due to a small amount of high-frequency dispersive radiation generated by the periodically-forced optical pulse, which at certain characteristic sideband frequencies is able to phase-match with the amplifiers (when their spacing is exactly periodic). In situations where the ratio of the amplifier spacing to the dispersion length is not sufficiently small, it was shown that this sideband radiation can grow with distance and eventually become large enough to affect the pulse dynamics.

The requirements for resolving this sideband radiation were shown to impose a number of restrictions on the use of a straightforward pseudo-spectral method. In particular, a large computational region is required in order to adequately resolve these sidebands in the frequency domain, which in turns implies that a large number of Fourier modes are needed to give adequate resolution in time. Other methods, such as a pseudo-spectral method employing fractional Fourier transforms [39,40], may relieve this restriction.

Finally, it should be noted that although the numerical computations must be done carefully in order to properly resolve the sideband radiation, the sensitivity of this dispersive radiation to various perturbations

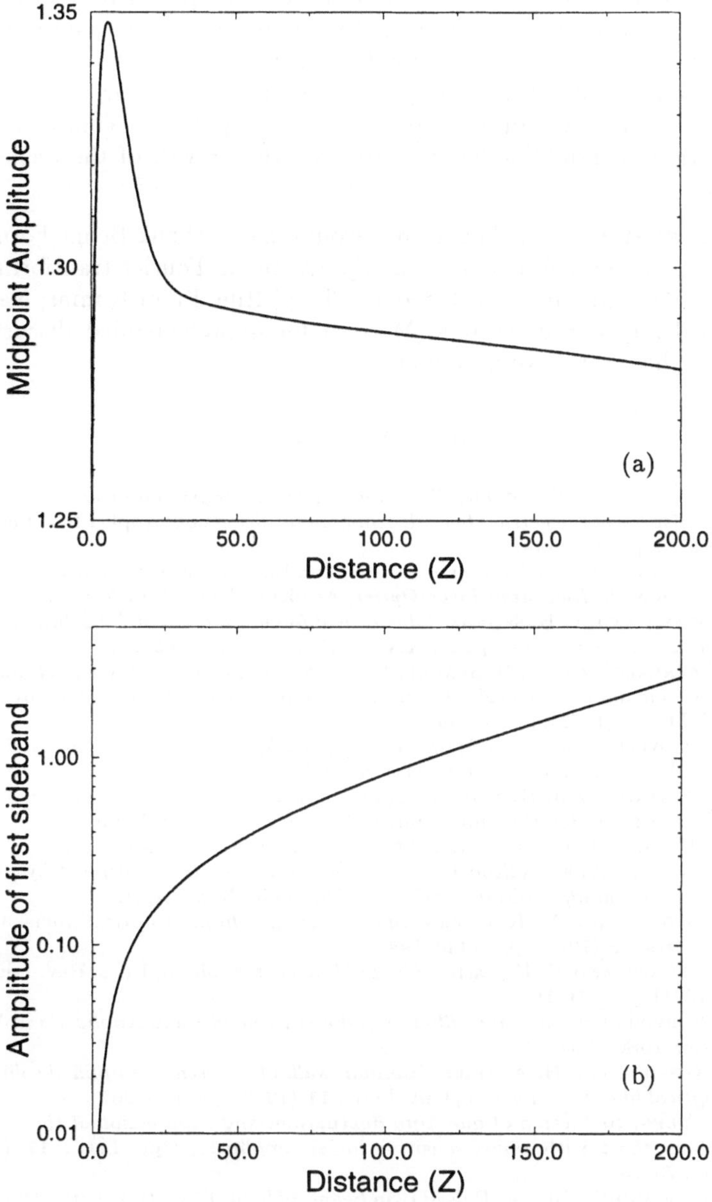

FIG. 4.4. *The height of the pulse midpoint in the time domain as a function of distance is plotted in (a), and the height of the peak of the first sideband in the frequency domain is shown in (b). Note that the roughly exponential growth of the sideband(s) causes the main pulse to decay with distance.*

(from inadequate computational resolution, for example, or from too much damping in the absorbing layers near the boundary of the computational region) is good from a practical point of view. In particular, this means that perturbations which are likely to occur in any realistic implementation of phase-sensitive amplifiers (such as the amplifier spacing being not exactly periodic) should suffice to eliminate the growth of the dispersive sidebands.

Acknowledgments. The author would like to thank Bengt Fornberg for pointing out the references about the fractional Fourier transform. He would also like to thank Chris Goedde, Cheryl Hile, Prem Kumar, Nathan Kutz, Ruo-Ding Li and Antonio Mecozzi for many extended discussions concerning phase-sensitive amplifiers.

REFERENCES

[1] A. Hasegawa and F. Tappert, *Transmission of stationary nonlinear optical pulses in dispersive dielectric fibers. I. Anomalous dispersion*, Appl. Phys. Lett., **23** (1973), pp. 142–144.
[2] A. Hasegawa, *Optical Solitons in Fibers*, 2nd Ed., Springer, Berlin, 1990.
[3] G. P. Agrawal, *Nonlinear Fiber Optics*, Academic Press, New York, 1989.
[4] Y. Kodama and A. Hasegawa, *Theoretical foundation of optical-soliton concept in fibers*, *Progress in Optics*, **XXX** #IV (1992), pp. 205–259.
[5] V. E. Zakharov and A. B. Shabat, *Exact theory of two-dimensional self-focusing and one-dimensional self-modulation of waves in nonlinear media*, Sov. Phys. JETP, **34** (1972), pp. 62–69.
[6] A. C. Newell, *Solitons in Mathematics and Physics*, Society for Industrial and Applied Mathematics, Philadelphia, 1985.
[7] M. J. Ablowitz and H. Segur, *Solitons and the Inverse Scattering Transform*, Society for Industrial and Applied Mathematics, Philadelphia, 1981.
[8] H. A. Haus, *Molding light into solitons*, IEEE Spectrum, March 1993.
[9] L. F. Mollenauer, *Soliton transmission speeds greatly multiplied by sliding-frequency guiding filters*, Optics and Photonics News, April 1994.
[10] A. Hasegawa and Y. Kodama, *Guiding-center soliton in optical fibers*, Optics Letters, **15** (1990), pp. 1443–1445.
[11] A. Hasegawa and Y. Kodama, *The guiding center soliton*, Phys. Rev. Lett., **66** (1991), pp. 161–164.
[12] E. Desurvire, *Erbium-doped fiber amplifiers: Principles and Applications*, Wiley, New York, 1994.
[13] J. P. Gordon and H. A. Haus, *Random walk of coherently amplified solitons in optical fiber transmission*, Opt. Lett., **11** (1986), pp. 665–667.
[14] H. P. Yuen, *Reduction of quantum fluctuations and suppression of the Gordon-Haus effect with phase-sensitive linear amplifiers*, Opt. Lett., **17** (1992), pp. 73–75.
[15] O. Aytür and P. Kumar, *Pulsed twin beams of light*, Phys. Rev. Lett., **13** (1990), pp. 1551–1554.
[16] W. Sohler and H. Suche, *Optical parametric amplification in Ti-diffused $LiNbO_3$ waveguides*, Appl. Phys. Lett., **37** (1980), pp. 255–257.
[17] S. Helmfrid, F. Laurell, and G. Arvidsson, *Optical parametric amplification of a 1.54 μm single-mode DFB laser in a Ti:$LiNbO_3$ waveguide*, J. Lightwave Technol., **11** (1990), pp. 1459–1469.
[18] M. E. Marhic and C.-H. Hsia, *Optical amplification in a nonlinear interferometer*, Elect. Lett., **27** (1991), pp. 210–211.

[19] G. BARTOLINI, R.-D. LI, P. KUMAR, W. RIHA AND K. V. REDDY, *1.5 μm phase-sensitive amplifier for ultra-high speed communications*, OFC'94 Technical Digest, Vol. 4, (Optical Society of America, Washington, D.C., 1994), pp. 202–203.

[20] J. N. KUTZ, W. L. KATH, R.-D. LI AND P. KUMAR, *Long-distance pulse propagation in nonlinear optical fibers using periodically-spaced parametric amplifiers*, Opt. Lett., **18** (1993), pp. 802-804.

[21] J. N. KUTZ, C. V. HILE, W. L. KATH, R.-D. LI AND P. KUMAR, *Pulse propagation in nonlinear optical fiber-lines that employ phase-sensitive parametric amplifiers*, J. Opt. Soc. Amer. B, **11** (1994), pp. 2112–2123.

[22] R.-D. LI, P. KUMAR, W. L. KATH AND J. N. KUTZ, *Combating dispersion with parametric amplifiers, IEEE Photonics Technology Letters*, **5** (1993), pp. 669–672.

[23] R.-D. LI, P. KUMAR, AND W. L. KATH, *Dispersion compensation with phase-sensitive amplifiers*, J. Lightwave Technology, **12** (1994), pp. 541–549.

[24] A. MECOZZI, P. KUMAR, W. L. KATH AND C. G. GOEDDE, *Long-term storage of a soliton bit stream using phase-sensitive amplification*, Opt. Lett., **19** (1994), pp. 2050-2052.

[25] C. G. GOEDDE, W. L. KATH AND P. KUMAR, *Compensation of the soliton self-frequency shift with phase-sensitive amplifiers*, Opt. Lett., **19** (1994), pp. 2077-2079.

[26] J. KEVORKIAN AND J. D. COLE, *Perturbation Methods in Applied Mathematics*, (Springer-Verlag, 1981).

[27] M. C. CROSS AND P. C. HOHENBERG, *Pattern formation outside of equilibrium*, Rev. Mod. Phys., **65** (1993), pp. 851–1112.

[28] J. N. KUTZ AND W. L. KATH, *Stability of pulses in nonlinear optical fibers using phase-sensitive amplifiers*, submitted to SIAM J. Applied Math.

[29] M. WEINSTEIN, *Modulational stability of ground states of nonlinear Schrödinger Equations*, SIAM J. Math. Anal., **16** (1985), pp. 472–491.

[30] B. FORNBERG AND G. B. WHITHAM, *A numerical and theoretical study of certain nonlinear wave phenomena*, Proc. Phil. Trans. Lond. A, **289** (1978), pp. 373–404.

[31] T. Y. HOU, J. S. LOWENGRUB, AND M. J. SHELLEY, *Removing the stiffness from interfacial flows with surface tension*, J. Comp. Phys., **114** (1994), pp. 312–338.

[32] M. D. FEIT AND J. A. FLECK, *Light propagation in graded-index optical fibers*, Appl. Opt., **17** (1978), pp. 3990–3998.

[33] E. LAGASSE AND R. BAETS, *Application of propagating beam methods to electromagnetic and acoustic wave propagation problems: A review*, Radio Science, **22** (1987), pp. 1225–1233.

[34] T. R. TAHA AND M. J. ABLOWITZ, *Analytical and numerical aspects of certain nonlinear evolution equations. II. Numerical, nonlinear Schrödinger equation*, J. Comp. Phys., **55** (1984), pp. 203–230.

[35] F. IF, P. BERG, P. L. CHRISTIANSEN AND O. SKOVGAARD, *Split-step spectral method for nonlinear Schrödinger-equation with absorbing boundaries*, J. Comp. Phys., **72** (1987), pp. 501–503.

[36] J. P. GORDON, *Dispersive perturbations of solitons of the nonlinear Schrödinger equation*, J. Opt. Soc. Am. B, **9** (1992), pp. 91–97.

[37] H. SEGUR, S. TANVEER AND H. LEVINE, EDS., *Asymptotics Beyond All Orders*, Plenum, New York, (1992).

[38] J. NATHAN KUTZ, Ph.D. Thesis, Northwestern University, 1994.

[39] D. H. BAILEY AND P. N. SWARZTRAUBER, *The fractional Fourier transform and applications*, SIAM Review, **33** (1991), pp. 389–404.

[40] D. H. BAILEY AND P. N. SWARZTRAUBER, *A fast method for the numerical evaluation of continuous Fourier and Laplace transforms*, SIAM J. Sci. Comput., **15** (1994), pp. 1105–1110.

NUMERICAL SOLUTION OF PROBLEMS WITH DIFFERENT TIME SCALES II

HEINZ-OTTO KREISS*

1. Introduction. Consider the initial value problem for a system of differential equations

$$(1.1)\qquad \begin{aligned} u_t &= \varepsilon^{-1}Au + Bu + F, \quad t \geq 0, \\ u(0) &= u_0. \end{aligned}$$

Here $u = (u_1, \ldots, u_n)^T$, $F = F(t,\varepsilon) = (F_1(t,\varepsilon), \ldots, F_n(t,\varepsilon))^T \in C^\infty(t,\varepsilon)$ are vector functions with n components and $A = A(t,\varepsilon) \in C^\infty(t,\varepsilon)$, $B = B(t,\varepsilon) \in C^\infty(t,\varepsilon)$ are $n \times n$ matrices which depend on t and a small parameter ε for $0 < \varepsilon \leq \varepsilon_0 << 1$. We assume also that A, B, F vary slowly as functions of t. To make this precise we define

Definition 1.1. *A function $f(t,\varepsilon)$ is said to vary slowly of order p if f and its derivatives $d^j f/dt^j$, $j \leq p$ can be estimated independently of ε. If such an estimate holds for any p, then we say that the function varies slowly.*

We are not interested in problems with rapidly growing solutions where the large matrix is degenerate like

$$u_t = \left(\frac{1}{\varepsilon} \begin{pmatrix} 0 & 1 \\ 0 & 0 \end{pmatrix} + \begin{pmatrix} b_{11} & b_{12} \\ 0 & b_{22} \end{pmatrix} \right) u.$$

Therefore, we make

Assumption 1.1. *There is a slowly varying transformation $S(t,\varepsilon)$ such that*

$$(1.2)\qquad S^{-1}(t,\varepsilon)A(t,\varepsilon)S(t,\varepsilon) = \begin{pmatrix} A_{11}(t,\varepsilon) & 0 \\ 0 & \varepsilon A_{22}(t,\varepsilon) \end{pmatrix},$$

where A_{11}, A_{22} are slowly varying and

$$(1.3)\qquad A_{11} = \begin{pmatrix} \lambda_1 & & 0 \\ & \ddots & \\ 0 & & \lambda_m \end{pmatrix}, \quad |\lambda_j| \geq 1, \quad j = 1, 2, \ldots, m.$$

In this paper we are only interested in the highly oscillatory case and, therefore, we make

* Department of Mathematics, UCLA, Los Angeles, CA 90024.

Assumption 1.2. *The eigenvalues of A_{11} are purely imaginary, i.e.,*

$$\operatorname{Re}\lambda_j = 0, \quad j = 1, 2, \ldots, m.$$

The real stiff case is treated in [2].

An important special case arises when $A_{22} \equiv 0$. To achieve this goal one has to be careful how one splits a given matrix into its large part $\frac{1}{\varepsilon}A$ and its small part B. An example is

$$u_t = \left(i \begin{pmatrix} -\frac{1}{\varepsilon} & \frac{1}{\varepsilon} \\ \frac{1}{\varepsilon} & -(\frac{1}{\varepsilon}+1) \end{pmatrix} + \begin{pmatrix} b_{11} & b_{12} \\ b_{21} & b_{22} \end{pmatrix} \right) u =: \left(\frac{1}{\varepsilon}A + B \right) u.$$

The elements of $\frac{1}{\varepsilon}A$ are all large but the rows are almost linearly dependent. The eigenvalues of $\frac{1}{\varepsilon}A$ are

$$\lambda_1 = -\frac{2i}{\varepsilon} + \mathcal{O}(1), \quad \lambda_2 = -\frac{i}{2} + \mathcal{O}(\varepsilon)$$

and, therefore, $A_{22} \neq 0$. The remedy is to write the system in the form

$$u_t = \left(i \begin{pmatrix} -\frac{1}{\varepsilon} & \frac{1}{\varepsilon} \\ \frac{1}{\varepsilon} & -\frac{1}{\varepsilon} \end{pmatrix} + \begin{pmatrix} b_{11} & b_{12} \\ b_{21} & b_{22} - i \end{pmatrix} \right) u.$$

There is one other situation we have to be aware of. Consider

$$u_t = \left(\begin{pmatrix} 0 & \frac{1}{\varepsilon^2} \\ -1 & 0 \end{pmatrix} + \begin{pmatrix} b_{11} & b_{12} \\ 0 & b_{22} \end{pmatrix} \right) u \tag{1.4}$$

which could also be written as

$$u_t = \left(\begin{pmatrix} 0 & \frac{1}{\varepsilon^2} \\ 0 & 0 \end{pmatrix} + \begin{pmatrix} b_{11} & b_{12} \\ -1 & b_{22} \end{pmatrix} \right) u.$$

Both versions violate our assumption. Instead, we have to "rescale" the system by introducing new dependent variables

$$\varepsilon u_1 = u_1', \quad u_2 = u_2'.$$

Now (1.4) becomes

$$u_t' = \left(\frac{1}{\varepsilon} \begin{pmatrix} 0 & 1 \\ -1 & 0 \end{pmatrix} + \begin{pmatrix} b_{11} & \varepsilon b_{12} \\ 0 & b_{22} \end{pmatrix} \right) u',$$

and our assumptions are satisfied. This behavior is very common when discretizing hyperbolic problems with different time scales.

The plan of our paper is as follows. In Section 2 we collect analytic results for the solution of (1.1). In Section 3 we shall solve our problem by asymptotic expansions. In this paper we are not interested in highly oscillatory solutions. Therefore, in Section 4 we shall discuss methods to "initialize the data" such that the resulting solution varies only on the slow time scale.

In Section 5 we will discuss the backward differentiation methods (BDF)

$$\left(I - \frac{2}{3}k\left(\frac{1}{\varepsilon}A_{n+1} + B_{n+1}\right)\right) y_{n+1} = \sum_{j=0}^{p-1} \alpha_j y_{n-j} + kF_{n+1}$$

and, in Section 6, apply them to hyperbolic partial differential equations.

In many applications it is much easier to invert $I - \frac{2}{3}\frac{k}{\varepsilon}A$ than $I - \frac{2}{3}k(\frac{1}{\varepsilon}A + B)$. Therefore, in the last two sections, we consider split methods

$$\left(I - \frac{2}{3}\frac{k}{\varepsilon}A_{n+1}\right) y_{n+1} = \sum_{j=0}^{p-1} \alpha_j y_{n-j} + \sum_{j=0}^{r-1} \beta_j (B_{n-j} y_{n-j} + F_{n-j}). \tag{1.5}$$

Our results indicate that the best methods are

1) The classic combination of Leap-frog and Crank-Nicolson.
2) A combination of the second order BDF method with an explicit second or third order method of type (1.5).

In the literature one can find other splitting techniques like "Strang" and additative splitting. In our opinion, they cannot compete with split multistep methods. (For details, see [1]).

2. Analytic results. We use Assumption 1.1 to introduce a new variable $u_1 = S^{-1}u$ into (1.1) and obtain

$$u_{1t} = \left(\begin{pmatrix} \frac{1}{\varepsilon}A_{11} & 0 \\ 0 & A_{22} \end{pmatrix} + B^{(1)}\right) u_1 + F^{(1)}, \tag{2.1}$$

where

$$B^{(1)} = S^{-1}BS - S^{-1}S_t, \quad F^{(1)} = S^{-1}F.$$

There is a slowly varying transformation $I + \varepsilon S_1$ such that

$$(I + \varepsilon S_1)^{-1} \left(\begin{pmatrix} \frac{1}{2}A_{11} & 0 \\ 0 & A_{22} \end{pmatrix} + B^{(1)}\right) (I + \varepsilon S_1)$$

$$= \begin{pmatrix} \frac{1}{\varepsilon}(A_{11} + \varepsilon C_{11}^{(1)}) & 0 \\ 0 & A_{22} + \varepsilon C_{22}^{(1)} \end{pmatrix} =: D.$$

Here $C_{11}^{(1)}$ is diagonal. Therefore, $u_2 = (I + \varepsilon S_1)^{-1} u_1$ satisfies

$$u_{2t} = (D + \varepsilon B^{(2)}) u_2 + F^{(2)}.$$

We can repeat the process and after $p+1$ transformations we obtain a system of the form

$$\begin{pmatrix} u^I \\ u^{II} \end{pmatrix}_t = \begin{pmatrix} \frac{1}{\varepsilon}(A_{11}+\varepsilon C_{11}) & \varepsilon^p C_{12} \\ \varepsilon^p C_{21} & B_{22}+\varepsilon C_{22} \end{pmatrix} \begin{pmatrix} u^I \\ u^{II} \end{pmatrix} + F_2.$$

Thus, except for terms of order ε^p, we obtain the decoupled system

$$u_t^I = \frac{1}{\varepsilon}(A_{11}+\varepsilon C_{11})u^I + F^I, \quad C_{11} \text{ diagonal}, \tag{2.2a}$$

$$u_t^{II} = (B_{22}+\varepsilon C_{22})u^{II} + F^{II}. \tag{2.2b}$$

We shall now derive an asymptotic expansion for the solutions of (2.2a). We want to show that it consists of a slowly varying and a rapidly varying part.

If we neglect u_t^I, then $u^I = -\varepsilon(A_{11}+\varepsilon C_{11})^{-1}F^I$. Therefore, we make the substitution

$$u^I = \varepsilon\varphi_1 + u_1^I, \quad \varphi_1 = -(A_{11}+\varepsilon C_{11})^{-1}F$$

and obtain

$$u_{1t}^I = \frac{1}{\varepsilon}(A_{11}+\varepsilon C_{11})u^I + \varepsilon F_1^I, \quad F_1^I = ((A_{11}+\varepsilon C_{11})^{-1}F)_t.$$

Thus, we have reduced the forcing to order $\mathcal{O}(\varepsilon)$. After p steps we obtain

$$u^I = \sum_{j=1}^{p} \varepsilon^j \varphi_j + u_p^I, \tag{2.3}$$

where

$$\begin{aligned} u_{pt}^I &= \frac{1}{\varepsilon}(A_{11}+\varepsilon C_{11})u_p^I + \varepsilon^p F_p^I, \\ u_p^I(0) &= u^I(0) - \sum_{j=1}^{p} \varepsilon^j \varphi_j(0). \end{aligned}$$

If the initial data are such that $u_p^I(0) \equiv \mathcal{O}(\varepsilon^p)$, then $u_p^I(t) = \mathcal{O}(\varepsilon^p)$ and has p derivatives bounded independently of ε. Otherwise, it is dominated by the rapidly varying part

$$u_p^I(t) = e^{\frac{1}{\varepsilon}\int_0^t (A_{11}+\varepsilon C_{11})d\xi}\, u_p^I(0). \tag{2.4}$$

Since the solutions of (2.2b) are slowly varying we have sketched a proof of (for more details, see[4],[5] and [6])

Theorem 2.1. *The solution of (1.1) can be written as*

$$u = u_S + u_F.$$

Here u_S is slowly and u_F is rapidly varying. Any initial data $u(0)$ can be split into

$$u(0) = u_S(0) + u_F(0).$$

$u_S(0)$ generates the slowly varying and $u_F(0)$ the rapidly varying part. In particular, $u_S(t)$ is determined by (2.2b) and (2.3) with $u_p^I \equiv 0$. Also, u_S is a smooth function of ε.

3. Numerical solution by asymptotic expansion. To use asymptotic expansions to solve (1.1) we need to calculate the transformation (1.3) efficiently. Techniques to do so are described in [5]. We assume that we have executed this transformation and that the equations are given in the form

$$u_t^I = \varepsilon^{-1}(A_{11} + \varepsilon B_{11})u^I + B_{12}u^{II} + F^I, \tag{3.1a}$$

$$u_t^{II} = (A_{22} + B_{22})u^{II} + B_{21}u^I + F^{II}. \tag{3.1b}$$

For slowly varying solutions, we can, to first approximation, neglect u_t^I. Therefore, we solve

$$\begin{aligned} &(A_{11} + \varepsilon B_{11})\psi_0^I + \varepsilon B_{12}\psi_0^{II} + \varepsilon F^I = 0, \\ &\psi_{0t}^{II} = (A_{22} + B_{22})\psi_0^{II} + B_{21}\psi_0^I + F^{II}, \\ &\psi_0^{II}(0) = u^{II}(0). \end{aligned} \tag{3.2}$$

The remainder

$$u_1 = u - \psi_0$$

is a solution of

$$\begin{aligned} &u_{1t}^I = \varepsilon^{-1}(A_{11} + \varepsilon B_{11})u_1^I + B_{12}u_1^{II} - \psi_{0t}^I, \\ &u_{1t}^{II} = (A_{22} + \varepsilon B_{22})u_1^{II} + B_{21}u_1^I, \\ &u_1^{II}(0) = 0. \end{aligned} \tag{3.3}$$

Now we solve

$$(A_{11} + \varepsilon B_{11})\psi_1^I + \varepsilon B_{12}\psi_1^{II} - \varepsilon^{-1}\psi_{0t}^I = 0,$$

$$\psi_{1t}^{II} = (A_{22} + B_{22})\psi_1^{II} + B_{21}\psi_1^I, \tag{3.4}$$

$$\psi_1^{II}(0) = 0.$$

(Observe that $\psi_{0t}^I = \mathcal{O}(\varepsilon)$ and, therefore, $\psi_1^I = \mathcal{O}(1)$.) Then

$$u_2 = u_1 - \varepsilon^2 \psi_1$$

is a solution of

$$u_{2t}^I = \varepsilon^{-1}(A_{11} + \varepsilon B_{11})u_2^I + B_{12}u_2^{II} - \varepsilon^2 \psi_{1t}^I,$$

$$u_{2t}^{II} = (A_{22} + \varepsilon B_{22})u_2^{II} + B_{21}u_2^I, \tag{3.5}$$

$$u_2^{II}(0) = 0.$$

The reduction process can be continued. After p steps we obtain

$$u = \psi^{(p)} + w, \quad \psi^{(p)} = \psi_0 + \varepsilon \sum_{j=1}^{p-1} \varepsilon^j \psi_j,$$

where, except for terms of order $\mathcal{O}(\varepsilon^{p+1})$, w solves

$$\begin{pmatrix} w^I \\ w^{II} \end{pmatrix}_t = \begin{pmatrix} \frac{1}{\varepsilon}(A_{11} + \varepsilon B_{11}) & B_{12} \\ B_{21} & A_{22} + B_{22} \end{pmatrix} \begin{pmatrix} w^I \\ w^{II} \end{pmatrix}$$

$$=: \begin{pmatrix} D_{11} & D_{12} \\ D_{21} & D_{22} \end{pmatrix} \begin{pmatrix} w^I \\ w^{II} \end{pmatrix} \tag{3.6}$$

$$w^I(0) = u^I(0) - \left(\psi^{(p)}(0)\right)^I, \quad w^{II}(0) = 0.$$

If $w^I(0) = 0$, then $w \equiv 0$ and we have solved our problem, which, in this case, has no fast part. If $w^I(0) \neq 0$, then we have to separate the scales further. We make a variable substitution of type

$$w = \begin{pmatrix} I & S \\ 0 & I \end{pmatrix} \tilde{w}$$

and obtain

$$\tilde{w}_t + \begin{pmatrix} 0 & S_t \\ 0 & 0 \end{pmatrix} \tilde{w} = \begin{pmatrix} I & -S \\ 0 & I \end{pmatrix} \begin{pmatrix} D_{11} & D_{12} \\ D_{21} & D_{22} \end{pmatrix} \begin{pmatrix} I & S \\ 0 & I \end{pmatrix} \tilde{w}$$

$$= \begin{pmatrix} D_{11} - SD_{21} & D_{12} - SD_{22} \\ D_{21} & D_{22} \end{pmatrix} \begin{pmatrix} I & S \\ 0 & I \end{pmatrix} \tilde{w}$$

$$= \begin{pmatrix} D_{11} - SD_{21} & D_{11}S - SD_{22} + D_{12} - SD_{21}S \\ D_{21} & D_{21}S + D_{22} \end{pmatrix} \tilde{w}.$$

Thus,

$$\begin{aligned} \tilde{w}_t &= \left(\begin{pmatrix} D_{11} - SD_{21} & 0 \\ D_{21} & D_{21}S + D_{22} \end{pmatrix} + \mathcal{O}(\varepsilon^{p+1}) \right) \tilde{w} \\ \tilde{w}^I(0) &= w^I(0), \quad \tilde{w}^{II}(0) = 0, \end{aligned} \tag{3.7}$$

if we choose S such that

$$\begin{aligned} S_t &= D_{11}S - SD_{22} + D_{12} - SD_{21}S + \mathcal{O}(\varepsilon^{p+1}) \\ &= \frac{1}{\varepsilon}(A_{11} + \varepsilon B_{11})S - S(A_{22} + B_{22}) + B_{12} - SB_{21}S + \mathcal{O}(\varepsilon^{p+1}). \end{aligned} \tag{3.8}$$

(3.8) is satisfied if we use $p+1$ steps of the iteration

$$\begin{aligned} \varepsilon S_t^{(j)} &= A_{11}S^{(j+1)} + \varepsilon B_{12} + \varepsilon B_{11}S^{(j)} - \varepsilon S^{(j)}(A_{22} + B_{22}) - \varepsilon S^{(j)} B_{21} S^{(j)} \\ S^{(1)} &= -\varepsilon A_{11}^{-1} B_{12}. \end{aligned}$$

Now we neglect the term $\mathcal{O}(\varepsilon^{p+1})$ in (3.7) and make the substitution

$$\tilde{w} = \begin{pmatrix} I & 0 \\ T & I \end{pmatrix} \tilde{\tilde{w}}.$$

In the same way as above, neglecting terms of order $\mathcal{O}(\varepsilon^{p+1})$, we obtain

$$\begin{aligned} \tilde{\tilde{w}}_t &= \begin{pmatrix} D_{11} - SD_{21} & 0 \\ 0 & D_{21}S + D_{22} \end{pmatrix} \tilde{\tilde{w}} \\ &= \begin{pmatrix} \frac{1}{\varepsilon}(A_{11} + \varepsilon B_{11}) - SB_{21} & 0 \\ 0 & A_{22} + B_{22} + B_{21}S \end{pmatrix} \tilde{\tilde{w}} \\ \tilde{\tilde{w}}^I(0) &= w^I(0), \quad \tilde{\tilde{w}}^{II}(0) = Tw^I(0), \end{aligned} \tag{3.9}$$

if we choose T such that

$$\begin{aligned} T_t &= -T(D_{11} - SD_{21}) + D_{21} + (D_{21}S + D_{22})T + \mathcal{O}(\varepsilon^{p+1}) \\ &= -T\left(\frac{1}{\varepsilon}(A_{11} + \varepsilon B_{11}) - SB_{21}\right) + B_{21} + (B_{21}S + A_{22} + B_{22})T + \mathcal{O}(\varepsilon^{p+1}). \end{aligned} \tag{3.10}$$

Again, we can determine T by iteration starting with

$$T^{(1)} = \varepsilon B_{21} A_{11}^{-1}. \tag{3.11}$$

As before, the system for $\tilde{\tilde{w}}^I$ can be solved by analytic techniques. To determine $\tilde{\tilde{w}}^{II}$ we can use a standard multistep or Runge-Kutta type method.

Except for terms of order $\mathcal{O}(\varepsilon^{p+1})$ we have separated the scales. The slow and the fast parts of the solution of (3.9) are obtained by choosing $\tilde{\tilde{w}}^I(0) = 0$ and $\tilde{\tilde{w}}^{II}(0) = 0$, respectively. Since

$$w = \begin{pmatrix} I & S \\ 0 & I \end{pmatrix} \begin{pmatrix} I & 0 \\ T & I \end{pmatrix} \tilde{\tilde{w}} = \begin{pmatrix} I+ST & S \\ T & I \end{pmatrix} \tilde{\tilde{w}},$$

we have, therefore,

$$(3.12)\qquad \begin{aligned} u_S^I &= (\psi^{(p)})^I + w_S^I + \mathcal{O}(\varepsilon^{p+1}) = (\psi^{(p)})^I + S\tilde{\tilde{w}}^{II} + \mathcal{O}(\varepsilon^{p+1}), \\ u_S^{II} &= (\psi^{(p)})^{II} + w_S^{II} + \mathcal{O}(\varepsilon^{p+1}) = (\psi^{(p)})^{II} + \tilde{\tilde{w}}^{II} + \mathcal{O}(\varepsilon^{p+1}), \end{aligned}$$

$$(3.13)\qquad \begin{aligned} u_F^I &= w_F^I + \mathcal{O}(\varepsilon^{p+1}) = (I+ST)\tilde{\tilde{w}}^I + \mathcal{O}(\varepsilon^{p+1}), \\ u_F^{II} &= w_F^{II} + \mathcal{O}(\varepsilon^{p+1}) = T\tilde{\tilde{w}}^I + \mathcal{O}(\varepsilon^{p+1}), \end{aligned}$$

Observe that $S = \mathcal{O}(\varepsilon)$, $T = \mathcal{O}(\varepsilon)$, $\tilde{\tilde{w}}^{II} = \mathcal{O}(\varepsilon)$ and, therefore,

$$u_S^I = (\psi^{(0)})^I + \mathcal{O}(\varepsilon^2), \quad u_S^{II} = (\psi^{(0)})^{II} + \varepsilon\varphi^{II} + \mathcal{O}(\varepsilon^2)$$

where

$$\varphi_t^{II} = (A_{22} + B_{22})\varphi^{II}, \quad \varphi^{II}(0) = B_{21}A_{11}^{-1}u^I(0).$$

4. Initialization. If the solution of our problem varies also on the fast scale, then we must either separate the scales as explained in the previous section or use a standard explicit multistep or Runge-Kutta method directly with a timestep $k << \varepsilon$.

In many applications one is not interested in the fast scale. Often, the energy contained in the fast scale solution is small and might also be introduced through observational errors. Therefore, one initializes the data, i.e., one changes the the initial data such that the resulting solution varies only on the slow scale. There are different method to accomplish this.

1) *Separation of scales.* We use the technique shown in the previous section to obtain initialized data $u_J^I(0)$, $u_J^{II}(0)$. The easiest way is to use

$$(4.1)\qquad \begin{aligned} u_J^I(0) &= \psi_0^I(0) \approx -\varepsilon A_{11}^{-1}(B_{12}u^{II}(0) + F^I(0)), \\ u_J^{II}(0) &= \psi_0^{II}(0) = u^{II}(0). \end{aligned}$$

If the fast scale is only introduced through observational errors, we commit an error of order $\mathcal{O}(\varepsilon^2)$. If the fast scale is "real", then we should use a more accurate initialization. Neglecting terms of order $\mathcal{O}(\varepsilon^2)$, we obtain

$$
\begin{aligned}
u_J^I(0) &= \psi_0^I(0), \\
u_J^{II}(0) &= \psi_0^{II}(0) + \tilde{\tilde{w}}^{II}(0) = u^{II}(0) + \varepsilon B_{21} A_{11}^{-1} u^I(0).
\end{aligned}
\tag{4.2}
$$

The term $\varepsilon B_{21} A_{11}^{-1} u^I(0)$ represents the effect of the fast scale on the slow scale solution.

2) *The bounded derivative principle.* We determine the initial data such that p time derivatives of the solution are bounded independently of ε at $t = 0$. Typically $p = 2$, and then it is equivalent with (4.1). For details, see [5].

3) *Richardson extrapolation.* Let $k > 0$ denote a timestep which defines gridpoints $t_\nu = \nu k$, $\nu = 0, 1, 2, \dots$ and gridfunctions $y_\nu = y(\nu k)$. We approximate (1.1) by

$$
\begin{aligned}
&\left(I - \frac{k}{\varepsilon} A\right) y(t_{n+1}) = (I + k\tilde{B}) y(t_n) + kF(t_n), \\
&y(0) = u(0).
\end{aligned}
\tag{4.3}
$$

Here $A = A(t_{n+1}, \varepsilon)$, $\tilde{B} = B(t_n, \varepsilon) + kB^{(1)}(t_n, \varepsilon, k)$ are smooth functions of all variables.

Remark: In actual applications we use $\tilde{B} = B$. We have added $B^{(1)}$ to simplify the discussion below.

Our aim is to prove

Theorem 4.1. *Let $u_S = u_S(t, \varepsilon)$ be the smooth part of the solution of (1.1). For any integer $p > 0$, the solution $y = y(t_n, k, \varepsilon)$ of (4.3) can be expanded into a series*

$$
y(t_n, k, \varepsilon) = u_S(t_n, k, \varepsilon) + kE + \mathcal{O}(k^{p+1} + \varepsilon^{p+1} + (\varepsilon/k)^n).
\tag{4.4}
$$

Here kE stands for a Taylor expansion

$$
kE = ke_1(t_n, \varepsilon) + \cdots + k^p e_p(t_n, \varepsilon),
\tag{4.5}
$$

where $e_j(t, \varepsilon)$ are smooth functions of all variables. Therefore, for all n with $(\varepsilon/k)^n << k^{p+1}$, we can use Richardson extrapolation to calculate u_S accurately to order $\mathcal{O}(k^p)$.

The proof consists of a number of steps.

a) We use the transformation (1.2) to make a change of variables $y_{n+1} = S_{n+1}\tilde{y}_{n+1}$. Then we obtain an approximation of the same type for (2.1). Therefore, we can assume that A already has the blockdiagonal form (1.2).

b) Corresponding to the previous section, we can construct a particular slowly varying solution such that we can neglect the forcing F in (4.3). We solve

$$A_{11}\varphi_0^I(t_{n+1} + \varepsilon B_{11}\varphi_0^I(t_n) + \varepsilon\tilde{B}_{12}\varphi_0^{II}(t_n) + \varepsilon F^I(t_n) = 0,$$

$$(4.6)\, (I - kA_{22})\varphi_0^{II}(t_{n+1}) = (I + k\tilde{B}_{22})\varphi_0^{II}(t_n) + k\tilde{B}_{21}\varphi_0^I(t_n) + F^{II}(t_n),$$

$$\varphi_0^{II}(0) = u^{II}(0).$$

(4.6) is a regular approximation of (3.2). Therefore, standard techniques show that there is a solution of the form

$$\varphi_0 = \psi_0 + kE_1 + \mathcal{O}(k^{p+1}), \tag{4.7}$$

where kE_1 represents a Taylor expansion of type (4.5). The remainder

$$y_1 = y - \varphi_0$$

is a solution of

$$\begin{aligned} \left(I - \frac{k}{\varepsilon}A_{11}\right) y_1^I(t_{n+1}) &= (I + k\tilde{B}_{11})y_1^I(t_n) + k\tilde{B}_{12}y_1^{II}(t_n) \\ &\quad -(\varphi_0^I(t_{n+1} - \varphi_0^I(t_n)), \\ (I - kA_{22})y_1^{II}(t_{n+1}) &= (I + k\tilde{B}_{22})y_1^{II}(t_n) + k\tilde{B}_{21}y_1^I(t_n), \\ y_1^{II}(0) &= 0, \end{aligned} \tag{4.8}$$

which is an approximation of (3.3). Observing that

$$\varphi_0^I(t_{n+1}) - \varphi_0^I(t_n) = \psi_{0t} + kE_1 + \mathcal{O}(k^{p+1}),$$

we repeat the construction and obtain

$$\varphi_1 = \psi_1 + kE_2 + \mathcal{O}(k^{p+1}).$$

This process can be continued and, therefore, we have proved

Lemma 4.1. *Corresponding to the procedure in the previous section, we can construct a solution*

$$\varphi^{(p)} = \psi^{(p)} + kE + \mathcal{O}(k^{p+1})$$

such that, neglecting terms of order $\mathcal{O}(\varepsilon^{p+1} + k^{p+1})$,

$$v = y - \varphi^{(p)}$$

solves the homogeneous system (4.8) which we write as

$$
\begin{aligned}
v(t_{n+1}) &= \begin{pmatrix} \left(I - \frac{k}{\varepsilon}A_{11}\right)^{-1}(I + k\tilde{B}_{11}) & k\left(I - \frac{k}{\varepsilon}A_{11}\right)^{-1}\tilde{B}_{12} \\ k(I - kA_{22})^{-1}\tilde{B}_{21} & (I - kA_{22})^{-1}(I + k\tilde{B}_{22}) \end{pmatrix} v(t_n) \\
&=: \begin{pmatrix} C_{11} & C_{12} \\ C_{21} & C_{22} \end{pmatrix} v(t_n).
\end{aligned}
$$

We now introduce a transformation

$$
v(t_n) = \begin{pmatrix} I & \tilde{S}_n \\ 0 & I \end{pmatrix} \tilde{v}(t_n)
$$

and obtain

$$
\begin{aligned}
\tilde{v}(t_{n+1}) &= \begin{pmatrix} I & -\tilde{S}_{n+1} \\ 0 & I \end{pmatrix} \begin{pmatrix} C_{11} & C_{12} \\ C_{21} & C_{22} \end{pmatrix} \begin{pmatrix} I & \tilde{S}_n \\ 0 & I \end{pmatrix} \tilde{v}(t_n) \\
&= \begin{pmatrix} C_{11} - \tilde{S}_{n+1}C_{21} & C_{12} - \tilde{S}_{n+1}C_{22} \\ C_{21} & C_{22} \end{pmatrix} \begin{pmatrix} I & \tilde{S}_n \\ 0 & I \end{pmatrix} \tilde{v}(t_n) \\
&= \begin{pmatrix} C_{11} - \tilde{S}_{n+1}C_{21} & \tilde{C}_{12} \\ C_{21} & C_{22} + C_{21}\tilde{S}_n \end{pmatrix} \tilde{v}(t_n).
\end{aligned}
$$

We choose $\tilde{S}_n$ such that

$$
\begin{aligned}
\tilde{C}_{12} &=: C_{11}\tilde{S}_n - \tilde{S}_{n+1}C_{21}\tilde{S}_n + C_{12} - \tilde{S}_{n+1}C_{22} \\
&= (C_{11} - I)\tilde{S}_n + C_{12} - \tilde{S}_{n+1}(C_{22} - I) + (\tilde{S}_n - \tilde{S}_{n+1}) + \tilde{S}_{n+1}C_{21}\tilde{S}_n \\
&= \mathcal{O}(k^{p+2}).
\end{aligned}
$$

We multiply the last equation by $\frac{\varepsilon}{k}\left(I - \frac{k}{\varepsilon}A_{11}\right)$. Since

$$
\begin{aligned}
\tilde{B}_{ij} &= B_{ij} + kB_{ij}^{(1)}, \\
C_{11} - I &= \frac{k}{\varepsilon}\left(I - \frac{k}{\varepsilon}A_{11}\right)^{-1}(A_{11} + \varepsilon\tilde{B}_{11}), \\
C_{12} &= k\left(I - \frac{k}{\varepsilon}A_{11}\right)^{-1}\tilde{B}_{12}, \qquad (4.9) \\
C_{22} - I &= k(A_{22} + B_{22}) + k^2\tilde{\tilde{B}}_{22}, \\
C_{21} &= kB_{21} + k^2\tilde{\tilde{B}}_{21},
\end{aligned}
$$

we obtain

$$(A_{11} + \varepsilon B_{11})\tilde{S}_n + \varepsilon\tilde{B}_{12} - (\varepsilon I - kA_{11})\tilde{S}_{n+1}(A_{22} + B_{22} + k\tilde{\tilde{B}}_{22})$$

$$+(\varepsilon I - kA_{11})\frac{\tilde{S}_{n+1} - \tilde{S}_n}{k} + (\varepsilon I - kA_{11})\tilde{S}_{n+1}(B_{21} + k\tilde{B}_{21})\tilde{S}_n \tag{4.10}$$

$$= \mathcal{O}(\varepsilon k^{p+1} + k^{p+2}).$$

(4.10) is a $\mathcal{O}(k)$ approximation of (3.8). By iteration, we obtain

Lemma 4.2. *There is a transformation*

$$\tilde{S} = S + \varepsilon kE + \mathcal{O}(k^{p+1})$$

such that, neglecting terms of order $\mathcal{O}(k^{p+1})$,

$$\tilde{v}(t_{n+1}) = \begin{pmatrix} C_{11} - \tilde{S}_{n+1}C_{21} & 0 \\ C_{21} & C_{22} + C_{21}\tilde{S}_n \end{pmatrix} \tilde{v}(t_n). \tag{4.11}$$

As in the previous section, we transform (4.11) to blockdiagonal form by a substitution

$$\tilde{v}(t_n) = \begin{pmatrix} I & 0 \\ \tilde{T}_n & I \end{pmatrix} \tilde{\tilde{v}}(t_n).$$

We obtain

$$\tilde{\tilde{v}}(t_{n+1}) = \begin{pmatrix} C_{11} - \tilde{S}_{n+1}C_{21} & 0 \\ \tilde{C}_{21} & C_{22} + C_{21}\tilde{S}_n \end{pmatrix} \tilde{\tilde{v}}(t_n),$$

where, by (4.9),

$$\begin{aligned}
\tilde{C}_{21} &= -\tilde{T}_{n+1}(C_{11} - \tilde{S}_{n+1}C_{21}) + C_{21} + C_{22}\tilde{T}_n + C_{21}\tilde{S}_n\tilde{T}_n \\
&= \tilde{T}_{n+1}(I - C_{11}) + C_{21} + (C_{22} - I)\tilde{T}_n - (\tilde{T}_{n+1} - \tilde{T}_n) \\
&\quad +\tilde{T}_{n+1}\tilde{S}_{n+1}C_{21} + C_{21}\tilde{S}_n\tilde{T}_n \\
&= -\frac{k}{\varepsilon}\tilde{T}_{n+1}(I - \frac{k}{\varepsilon}A_{11})^{-1}(A_{11} + \varepsilon\tilde{B}_{11}) + k\tilde{B}_{21} + k^2\tilde{\tilde{B}}_{21} \\
&\quad +\big(k(A_{22} + \tilde{B}_{22}) + k^2\tilde{\tilde{B}}_{22}\big)\tilde{T}_n - \frac{\tilde{T}_{n+1} - \tilde{T}_n}{k} \\
&\quad +k\tilde{T}_{n+1}\tilde{S}_{n+1}(\tilde{B}_{21} + k\tilde{\tilde{B}}_{21}) + k(\tilde{B}_{21} + k\tilde{\tilde{B}}_{21})\tilde{S}_n\tilde{T}_n.
\end{aligned}$$

We choose $\tilde{T}$ such that $\tilde{C}_{21} = \mathcal{O}(k^{p+2})$. We divide the above relation by k. Observing that $\frac{1}{\varepsilon}\left(I - \frac{k}{\varepsilon}A_{11}\right)^{-1} = (\varepsilon I - kA_{11})^{-1}$ and, neglecting all terms of order $\mathcal{O}(k)$, we obtain

$$\begin{aligned} &-\tilde{T}_{n+1}(\varepsilon I - kA_{11})^{-1}(A_{11} + \varepsilon B_{11}) + B_{21} + (A_{22} + B_{22})\tilde{T}_{n+1} \\ &+\frac{\tilde{T}_{n+1} - \tilde{T}_n}{k} + \tilde{T}_{n+1}\tilde{S}_{n+1}B_{21} + B_{21}S_n\tilde{T}_n = \mathcal{O}(k^{p+1}). \end{aligned} \tag{4.12}$$

The last relation is a first order accurate approximation of (3.10). We solve it by iteration, starting with

$$\tilde{T}^{(1)} = B_{21}(A_{11} + \varepsilon B_{11})^{-1}(\varepsilon I - kA_{11}) = T + \mathcal{O}(k).$$

We have proved

Lemma 4.3. *There is a transformation*

$$\tilde{T} = T + kE + \mathcal{O}(k^{p+1})$$

such that, neglecting terms of order $\mathcal{O}(k^{p+1})$,

$$\tilde{\tilde{v}}(t_{n+1}) = \begin{pmatrix} C_{11} - \tilde{S}_{n+1}C_{21} & 0 \\ 0 & C_{22} + C_{21}\tilde{S}_n \end{pmatrix} \tilde{\tilde{v}}(t_n).$$

Since

$$|C_{11} - \tilde{S}_{n+1}C_{21}| = \mathcal{O}(\varepsilon/k + k\varepsilon),$$

the first components $\tilde{\tilde{v}}^I(t_n)$ decay rapidly. Also,

$$C_{22} + C_{21}\tilde{S}_n = I + k(A_{22} + B_{22} + B_{21}S) + \mathcal{O}(k^2)$$

and, therefore, by (3.9), there is an expansion

$$\tilde{\tilde{v}}^{II} = \tilde{\tilde{w}} + kE + \mathcal{O}(k^{p+1}).$$

Thus, we have proved Theorem 4.1.

The results in this section show that we can always initialize the data effectively and, therefore, we shall in the following sections only consider slowly varying solutions.

Remark: One could use Richardson extrapolation not only to initialize the data, i.e., use it for a limited number of timesteps, but as a numerical method for all times. However, there are stability problems, especially when A, B do not commute.

5. Backward differentiation methods. Let $k > 0$ denote the time step which defines gridpoints $t_\nu = \nu k$ and gridfunctions $y_\nu = y(\nu k)$. We are interested in the backward differentiation formulas

$$\left(I - k\alpha_{-1}\left(\frac{1}{\varepsilon}A + B\right)\right) y_{n+1} = \sum_{j=0}^{p-1} \alpha_j y_{n-j} + k\alpha_{-1}F_{n+1}. \tag{5.1}$$

Here A, B are evaluated at $t_{n+1} = (n+1)k$. Let u be a slowly varying solution of (1.1). Taylor expansion gives us

$$u_{n-j} = u_{n+1} - (j+1)ku'_{n+1} + \frac{(j+1)^2}{2}u''_{n+1} + \dots . \tag{5.2}$$

Therefore,

$$L[u] = \left(I - k\alpha_{-1}\left(\frac{1}{\varepsilon}A_{11} + B_{11}\right)\right) u_{n+1} - \sum_{j=0}^{p-1} \alpha_j u_{n-j} - k\alpha_{-1}F_{n+1}$$

$$= k\left\{\left(\sum_{j=0}^{p-1}(j+1)\alpha_j\right) u'_{n+1} - \alpha_{-1}\left(\frac{1}{\varepsilon}A + B\right) u_{n+1} - \alpha_{-1}F_{n+1}\right\}$$

$$\left(1 - \sum_{j=0}^{p-1}\alpha_j\right) u_{n+1} - \sum_{l=2}^{p} k^l \left(\sum_{j=0}^{p-1}\frac{(j+1)^l}{l!}\alpha_j\right)(-1)^l \frac{d^l u_{n+1}}{dt^l} + k^{p+1}R_{n+1}^{(p+1)}, \tag{5.3}$$

where

$$R^{(p+1)} = c_1\frac{d^{p+1}u}{dt^{p+1}} + kc_2\frac{d^{p+2}u}{dt^{p+2}} + \mathcal{O}\left(k^2\frac{d^{p+3}u}{dt^{p+3}}\right)$$

with

$$c_\nu = \frac{(-1)^{p+\nu}}{(p+\nu)!}\sum_{j=0}^{p-1}(j+1)^{p+\nu}\alpha_j, \quad \nu = 1, 2.$$

We choose

$$\sum_{j=0}^{p-1}\alpha_j = 1, \quad \sum_{j=0}^{p-1}(j+1)\alpha_j = \alpha_{-1}, \quad \sum_{j=0}^{p-1}(j+1)^l\alpha_j = 0, \quad l = 2, 3, \dots, p. \tag{5.4}$$

Then

$$L[u] = k^{p+1}R^{(p+1)} \tag{5.5}$$

and (5.1) is accurate of order p for slowly varying solutions. The methods are the usual BDF methods and the coefficients can be found in [3].

To obtain useful error estimates the method needs to be stable. For the stability investigation we assume that A, B are constant matrices. In the same way as for the continuous problem, we can separate the slow and the fast scales by transformation. Therefore, we need to investigate stability only for the decoupled systems (2.2a) and (2.2b). The stability question is governed by the stability region of the method when applied to the scalar differential equation

$$y' = \lambda y \tag{5.6}$$

with constant coefficients. We want to use our results also for the solution of hyperbolic partial differential equations, i.e., we want to apply it to systems of ordinary differential equations obtained by discretizing the space operators. As we have shown in [7], a desirable property of the multistep method is that it is locally stable, defined by

Definition 5.1. *A multistep method is called locally stable if the stability region Ω in the complex $\mu = k\lambda$ plane contains a halfdisc*

$$|\mu| = |k\lambda| \le R, \quad \operatorname{Re} k\lambda \le 0.$$

See Figure 5.1.

For $p = 1$ the scheme (5.1) represents the implicit Euler method which is A-stable and therefore also locally stable. The same is true for $p = 2$.

We shall discuss the stability region for the other values of p in a neighborhood of $\mu = 0$. The solutions of the difference approximation (5.1) are of the form

$$y_n = \kappa^n y_0, \tag{5.7}$$

where κ satisfies

$$(1 - \alpha_{-1}\lambda k)\kappa^p - \sum_{j=0}^{p-1} \alpha_j \kappa^{p-j-1} = 0. \tag{5.8}$$

The basic characteristic equation

$$\kappa^p - \sum_{j=0}^{p-1} \alpha_j \kappa^{p-j-1} = 0 \tag{5.9}$$

has exactly one root $\kappa = 1$ while all other roots satisfy

$$|\kappa_j| < 1, \quad j = 2, 3, \ldots, p.$$

Therefore, in a neighborhood of $\mu = 0$ we need only consider solutions of (5.8) which are of the form

$$\kappa = e^{\lambda k} + k^{p+1}\tau. \tag{5.10}$$

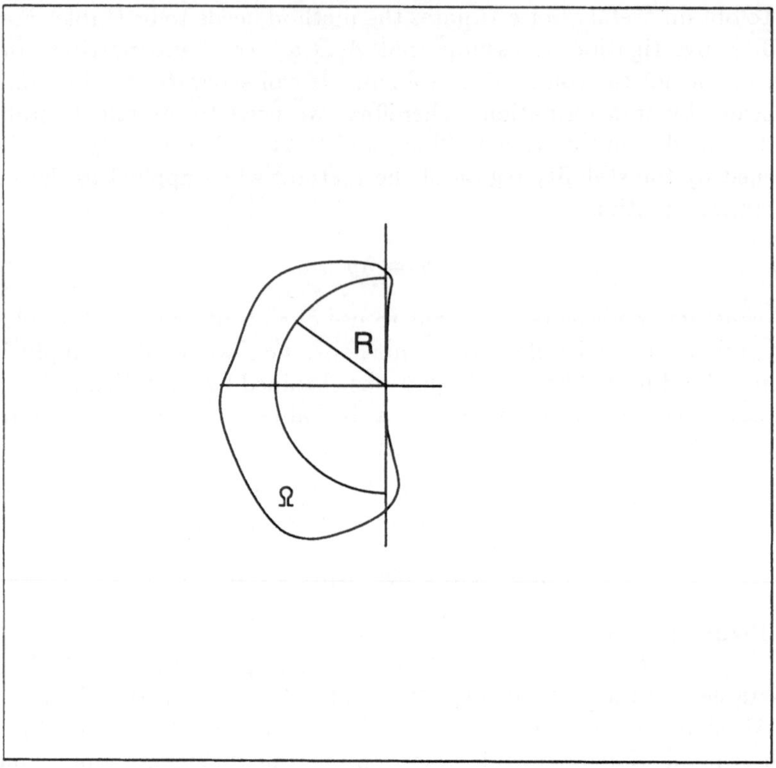

FIG. 5.1.

Introducing (5.10) into (5.8) gives us

$$(1-\alpha_{-1}\lambda k)(e^{\lambda k}+k^{p+1}\tau)^p-\sum_{j=0}^{p-1}\alpha_j(e^{\lambda k}+k^{p+1}\tau)^{p-j-1}=0.$$

Thus, we obtain the linearized equation

$$\left(p(1-\alpha_{-1}\lambda k)(e^{(p-1)\lambda k})-\sum_{j=0}^{p-1}\alpha_j e^{(p-j-2)\lambda k}(p-j-1)\right)k^{p+1}\tau=-\tilde{R} \tag{5.11}$$

where

$$\tilde{R}=(1-\alpha_{-1}\lambda k)e^{\lambda pk}-\sum_{j=0}^{p-1}\alpha_j e^{(p-j-1)\lambda k}.$$

By (5.6) and (5.3), truncation error analysis shows that

$$\begin{aligned} R_{n+1}^{(p+1)} &= L[e^{\lambda t}] = \left((1-\alpha_{-1}\lambda k)e^{\lambda pk} - \sum_{j=0}^{p-1} \alpha_j e^{(p-j-1)\lambda k}\right) e^{(n+1-p)\lambda k} \\ &= \tilde{R} \cdot e^{(n+1-p)\lambda k}. \end{aligned}$$

Therefore, from (5.11) we obtain, using Taylor expansion and (5.4),

(5.12) $\left(\alpha_{-1} + d_0\lambda k + \mathcal{O}((\lambda k)^2)\right) \tau = c_1\lambda^{p+1} + d_1\lambda^{p+2}k + \mathcal{O}(\lambda^{p+3}k^2).$

Here d_0, d_1 are real constants which can easily be calculated. The last relation finally gives us

$$\begin{aligned} \kappa &= e^{\lambda k} + k^{p+1}\tau \\ &= e^{\lambda k} + (\lambda k)^{p+1} \frac{c_1 + d_1\lambda k}{\alpha_{-1} + d_0\lambda k} + \mathcal{O}\left((\lambda k)^{p+3}\right) \\ &=: e^{\lambda k} + a_1(\lambda k)^{p+1} + a_2(\lambda k)^{p+2} + \mathcal{O}\left((\lambda k)^{p+3}\right). \end{aligned}$$

We now assume that $\lambda k = i\mu$ is purely imaginary. Then We now assume that $\lambda k = i\mu$ is purely imaginary. Then

(5.13) $$|\kappa|^2 = \begin{cases} 1 + 2a_1(-1)^{\frac{p+1}{2}} |\mu|^{p+1} + \mathcal{O}(|\mu|^{p+2}) & \text{if } p \equiv 1(2), \\ 1 - 2(a_1 - a_2)(-1)^{\frac{p+2}{2}} |\mu|^{p+2} + \mathcal{O}(|\mu|^{p+3}) & \text{if } p \equiv 0(2). \end{cases}$$

Thus, an interval on the imaginary axis belongs to Ω if

(5.13a) $$\begin{aligned} (-1)^{\frac{p+1}{2}} a_1 < 0 \quad &\text{for} \quad p \equiv 1(2), \\ (-1)^{\frac{p+2}{2}} (a_1 - a_2) > 0 \quad &\text{for} \quad p \equiv 0(2). \end{aligned}$$

In this case the method is locally stable. Simple calculations show that the method is not locally stable for $p = 3, 4$ but it is locally stable for $p = 5, 6$. See, for example, [3].

For $p = 3, 4, 5, 6$ we shall discuss another property of the stability region. Assume that

(5.14) $\lambda k = i\mu - \tau\mu^l, \quad l = 2\left[\frac{p+1}{2}\right], \quad ([x] \text{ largest integer } \leq x),$

where $\tau > 0$ is a constant which we choose later. In this case (5.13) becomes

$$|\kappa|^2 = \begin{cases} 1 - 2\left(\tau - a_1(-1)^{\frac{p+1}{2}}\right) \mu^{p+1} + \mathcal{O}(\mu^{p+2}) & \text{if } p \equiv 1(2), \\ 1 - 2\left(\tau + (a_1 - a_2)(-1)^{\frac{p+2}{2}}\right) \mu^{p+2} + \mathcal{O}(\mu^{p+3}) & \text{if } p \equiv 0(2). \end{cases}$$

(5.15)

Thus, the stability region Ω always contains a region $\Omega_0 \cap \Omega_1$, $\Omega_1 \cap \Omega_0$, where

$$\Omega_0 : \left\{ \text{Re } \lambda k \leq 0, \quad |\text{Re } \lambda k| \geq \tau |\text{Im } \lambda k|^l, \quad l = 2\left[\frac{p+1}{2}\right], \right. \tag{5.16}$$

$$\Omega_1 : \{|\lambda k| \leq R,$$

provided $R > 0$ is sufficiently small and τ sufficiently large. ($\tau = 0$ for $p = 5, 6$.)

The usual calculations show that, for sufficiently large τ, the whole region $\Omega_0 \in \Omega$. In Table 5.1 the minimum values of τ are calculated.

TABLE 5.1

p	3	4	5	6
τ	0.22	0.3	0.3	190

It shows that, for $p = 6$, τ has to be very large and the method is not useful.

We want to express the relation (5.16) also in another way. We assume that $A + B$ is blockdiagonal and consider the slow part

$$(I - k(A_{22} + B_{22}))\, y_{n+1}^{II} = \sum_{j=0}^{p-1} \alpha_j y_{n-j}^{II}. \tag{5.17}$$

We shall prove

Theorem 5.1. *Consider (5.17) for sufficiently small $k(|A_{22}| + |B_{22}|)$. We can stabilize (5.17) by changing $A_{22} + B_{22}$ to $A_{22} + B_{22} - \sigma\big(k(|A_{22}| + |B_{22}|)\big)^l I$, $l = 2[(p+1)/2]$, provided σ is sufficiently large.*

Proof. The eigenvalues λk are changed to $\lambda k - \sigma\,(k(|A_{22}| + |B_{22}|))^l$. Observing that $|\text{Im } \lambda k| \leq k(|A_{22}| + |B_{22}|)$ the theorem follows from (5.16).

6. Applications. Consider the scalar wave equation

$$w_t = a w_x, \quad 0 \leq x \leq 2\pi, \qquad w(x, 0) = f, \tag{6.1}$$

with periodic initial and boundary conditions. We discretize space by introducing gridpoints $x_\nu = \nu h$, $\nu = 0, \pm 1, \pm 2, \ldots$; $h = 2\pi/N$, and gridfunctions $u_\nu(t) = u(x_\nu, t)$ and approximate (6.1) by the dissipative approximation

$$u_t = aD_0u - \sigma h^3 D_+^2 D_-^2 u, \quad u = u_\nu, \qquad u(0) = f. \tag{6.2}$$

Here D_0, D_+, D_- are the usual centered, forward and backward difference operators, respectively. After Fourier transform (6.2) becomes

$$\hat{u}_t = \left(a\frac{i\sin\omega h}{h} - \frac{16\sigma}{h}\sin^4\frac{\omega h}{2}\right)\hat{u} =: \hat{Q}\hat{u}. \tag{6.3}$$

For time integration we choose the BDF methods. For $p = 1, 2$, the approximations are unconditionally stable. For $p = 3, 4$, we have to choose the timestep k and the dissipative coefficient σ such that $\hat{Q}$ belongs to the stability region. By (4.16) and Table 5.1, we need that

$$k\frac{16\sigma}{h}\sin^4\frac{\omega h}{2} \geq \frac{\tau k^4 \sin^4 \omega h}{h^4}a^4,$$

for all ωh, i.e.,

$$\sigma\frac{k}{h} \geq \tau\left(\frac{ka}{h}\right)^4 \quad \text{or} \quad \sigma \geq \tau|a|\left(\frac{k|a|}{h}\right)^3. \tag{6.4}$$

Thus, the amount of dissipation becomes rather large if we want to beat the speed of propagation, i.e., choose the timestep k such that $k|a|/h >> 1$.

For $p = 5, 6$, we can also choose τ such that (6.4) holds. For $p = 6$, we need to choose τ very large and, therefore, the method is not useful. Since for $p = 5, 6$, the schemes are locally stable, we can use $\tau = 0$, but then

$$|a\frac{k}{h}| \leq R. \tag{6.5}$$

The restriction (6.5) is more severe than the stability restriction for a good explicit method and, therefore, we shall not pursue this possibility further.

Now consider a hyperbolic system

$$w_t = Aw_x, \quad 0 \leq x \leq 2\pi, \qquad w(x, 0) = f, \tag{6.6}$$

and approximate it by

$$u_t = AD_0u - \sigma h^3 D_+ D_- u. \tag{6.7}$$

Let $a_1, \dots, a_n$ denote the eigenvalues of A. By (6.4) we need to choose

$$\sigma \ge \tau \max_j |a_j| \left(\frac{k \ \max_j |a_j|}{h} \right)^3 . \tag{6.8}$$

If all the a_j are of the same order of magnitude, this is not a severe restriction. However, for problems with different time scales, they are of different orders of magnitude. Then (6.7) is useless because the dissipation becomes too large. We can replace (6.7) by

$$u_t = AD_0 u - \sigma A^4 h^3 D_+ D_- u.$$

We transform the system to diagonal form and obtain

$$u_t = aD_0 u - \sigma a^4 h^3 D_+ D_- u, \quad a = a_j.$$

Our condition (6.8) now becomes

$$\sigma \ge \tau \left(\frac{k}{h} \right)^3 . \tag{6.9}$$

(6.9) is satisfactory. However, it is expensive to calculate the dissipation term. This is particularly true for problems in more than one space dimension. Therefore, we believe that only the first and second order BDF methods are useful to solve hyperbolic partial differential equations with different time scales.

7. Examples of semi-implicit methods. In many applications it is much easier to calculate $(I - \alpha_{-1} \frac{k}{\varepsilon} A)^{-1}$ than $(I - k\alpha_{-1}(\frac{1}{\varepsilon} A + B))^{-1}$. We can still use the backward differentiation formulas but we solve them by iteration.

$$\left(I - \alpha_{-1} \frac{k}{\varepsilon} A_{n+1} \right) y_{n+1}^{[j+1]} = \sum_{j=0}^{p-1} \alpha_j y_{n-j} + \alpha_{-1} k B_{n+1} y_{n+1}^{[j]}, \quad y^{[0]} \equiv 0. \tag{7.1}$$

(For simplicity only, we assume that $F \equiv 0$.)

Now assume that $A = -A^*$, $B = -B^*$ are skew-Hermitean. Without restriction we can also assume that A is blockdiagonal.

We want to investigate how many iterations we need to make such that the resulting method is stable. As we have seen in the previous section, only the first and second order BDF methods are useful for our purposes. We consider the second order method. Since there are no stability problems with the large part, we need only consider

$$\left(I - \frac{2k}{3} A \right) y_{n+1}^{[j+1]} = \frac{4}{3} y_n - \frac{1}{3} y_{n-1} + \frac{2}{3} k B y_{n+1}^{[j]}, \quad y^{[0]} \equiv 0.$$

Here we have deleted the index 22. We stop after s iterations and obtain

$$\begin{aligned} y_{n+1} &= \left(I - \frac{2k}{3}A\right)^{-1} \sum_{j=0}^{s-1} \left(\frac{2k}{3}B\left(I - \frac{2k}{3}A\right)^{-1}\right)^j \left(\frac{4}{3}y_n - \frac{1}{3}y_{n-1}\right) \\ &= \left(I - \frac{2k}{3}A\right)^{-1} \left(I - \frac{2k}{3}B\left(I - \frac{2k}{3}A\right)^{-1}\right)^{-1} \\ &\qquad \left(I - \left(\frac{2k}{3}B\left(I - \frac{2k}{3}A\right)^{-1}\right)^s\right) \left(\frac{4}{3}y_n - \frac{1}{3}y_{n-1}\right) \\ &= \left(I - \frac{2k}{3}(A+B)\right)^{-1} \left(I - \left(\frac{2k}{3}B\right)^s + \mathcal{O}\left(k^{s+1}|B|^s(|B| + |A|)\right)\right) \\ &\qquad \left(\frac{4}{3}y_n - \frac{1}{3}y_{n-1}\right). \end{aligned}$$

The worst stability situation occurs when $A = -B$. Then, to first approximation,

$$y_{n+1} = \left(I - \left(\frac{2k}{3}B\right)^s\right) \left((\frac{4}{3}y_n - \frac{1}{3}y_{n-1}\right).$$

Clearly, the approximation is not stable for $s = 1, 2$. For $s > 2$, we can write the approximation as

$$(I - kC)y_{n+1} = \frac{4}{3}y_n - \frac{1}{3}y_{n-1},$$

where

$$kC = \frac{2k}{3}(A+B) + \left(\frac{2k}{3}B\right)^s + \mathcal{O}\left(k^{s+1}|B|^s(|B| + |A|)\right).$$

The approximation is stable if $C + C^* \geq 0$. For $s \geq 3$ and $k(|A| + |B|)$ sufficiently small, we can, corresponding to the previous section, always stabilize the method by changing $A + B$ to $A + B - \sigma k^4 I$. Here $\sigma =$ const. $(|A^4| + |B^4|)$. If $s = 4$, we do not need the extra stabilization if B is nonsingular. The disadvantage of the method is that we need to invert the operator $I - \frac{2k}{3}A$ at least three times.

Another approach is to split the operator and use a mixed implicit-explicit approximation. The classic method is a combination of the Leap-frog and Crank-Nicholson methods.

$$\left(I - \frac{k}{\varepsilon}A_{n+1}\right) y_{n+1} = \left(I + \frac{k}{\varepsilon}A_{n-1}\right) y_{n-1} + 2kB_n y_n. \tag{7.2}$$

It has been used for a long time and is second order accurate for slowly varying solutions and stable if $A + A^* \leq 0$, $B = -B^*$, $|Bk| \leq 1$. As in the previous section, one needs a small timestep to calculate the solutions on the fast scale.

Let us discuss the behavior of the method with respect to the fast scale when $k >> \varepsilon$. Without restriction we can assume that A already has the blockdiagonal form (1.3). Then, neglecting terms of order $\mathcal{O}(\varepsilon)$, the equation for the fast scale becomes

$$y_{n+1}^I = \left(I - \frac{k}{\varepsilon}A_{11}\right)^{-1}\left(I + \frac{k}{\varepsilon}A_{11}\right) y_{n-1}^I,$$

and, therefore, for $k >> \varepsilon$,

$$y_{n+1}^I \simeq -y_{n-1}^I. \tag{7.3}$$

Thus, the fast scale behaves like a ± 1-wave and will not be rapidly damped.

If we have initialized the data properly, then the fast scale solution is not present and waves are not excited. However, if the initialization has not been done carefully or if fast waves are generated by, for example, rough data, then we will have difficulties. The remedy is to use time filters (see, for example, [8]) or use the Richardson extrapolation of Section 4 to re-initialize the solution periodically. To avoid this problem altogether one often uses a combination of the backward Euler and Leap-frog methods. Again, assuming that A is blockdiagonal, the scheme is given by

$$\left(I - \frac{2k}{\varepsilon}A_{11}\right) y_{n+1}^I = y_{n-1}^I + 2k(B_{11}y_n^I + B_{12}y_n^{II}), \tag{7.4a}$$

$$(I - 2kA_{22})y_{n+1}^{II} = y_{n-1}^{II} + 2k(B_{21}y_n^I + B_{22}y_n^{II}). \tag{7.4b}$$

As in Section 4, we can derive an asymptotic expansion and obtain, for y^I,

$$y_n^I = \tilde{y}_n^I + \varphi_0^I(t_n), \tag{7.5a}$$

where

$$\begin{aligned} \varphi_0^I(t_{n+1}) &= -\varepsilon A_{11}^{-1}B_{12}y_n^{II} + \mathcal{O}(\varepsilon^2), \\ \left(I - \frac{2k}{\varepsilon}A_{11}\right)\tilde{y}_{n+1}^I &= \tilde{y}_{n-1}^I. \end{aligned}$$

Now the rapidly varying part is decaying quickly. Introducing (7.5a) into (7.4b), neglecting terms of order $\mathcal{O}(k\varepsilon^2)$, gives us

$$(I - 2kA_{22})y_{n+1}^{II} = (I + 2k\varepsilon\tilde{B}_{22})y_{n-1}^{II} + 2k(B_{21}\tilde{y}_n^I + B_{22}y_n^{II}), \tag{7.5b}$$

where

$$\tilde{B}_{22} = -B_{21}A_{11}^{-1}B_{12}.$$

Now we will discuss the accuracy and the stability of (7.5b) for slowly varying solutions ($y_n^I \equiv 0$). If $A_{22} \neq 0$, then (7.5b) is only a first order approximation of (1.1). Also, we cannot expect it to be stable. To show this we apply it to a scalar problem:

$$A_{22} = ia, \quad B_{21} = 0, \quad B_{22} = ib,$$

i.e.,

$$(1 - 2ika)y_{n+1} = y_{n-1} + 2ikby_n.$$

We construct solutions of the form

$$y_n = \kappa^n y_0,$$

where κ are the solutions of the characteristic equation

$$(1 - 2ika)\kappa^2 - 2ikb\kappa - 1 = 0,$$

i.e.,

$$\begin{aligned}\kappa &= \frac{1}{1-2ika}\left(kbi \pm \sqrt{1 - 2ika - k^2b^2}\right)\\ &= \frac{1}{1-2ika}\left(ikb \pm \left(1 - ika - \frac{k^2b^2}{2} + \frac{k^2a^2}{2} + \mathcal{O}(k^3)\right)\right).\end{aligned}$$

Therefore,

$$\begin{aligned}|\kappa^2| &= \frac{1}{1+4k^2a^2}\left(1 + k^2(b \mp a)^2 - k^2b^2 + k^2a^2\right) + \mathcal{O}(k^4)\\ &= \frac{1 \mp 2k^2ab + 2k^2a^2}{1+4k^2a^2} + \mathcal{O}(k^4).\end{aligned}$$

If $ab \neq 0$ and $|b| > |a|$, then there is a k_0 such that $|\kappa| > 1$ for $0 < k < k_0$. Thus, the method is not locally stable.

Remark: One can prove the following theorem which is important for reaction-diffusion equations.

Theorem 7.1. *Assume that $A_{11} = A_{11}^* \leq 0$, $\tilde{B}_{22} = \tilde{B}_{22}^* \leq 0$, $B_{22} = -B_{22}^*$. Then the method (7.5b) is stable if $k(|B_{22}| + \varepsilon|\tilde{B}_{22}|) < 1$.* If $A_{22} \equiv 0$, then the method is, formally, still first order, but since $\varepsilon\tilde{B}_{22}$ is of order $\mathcal{O}(\varepsilon)$, the error is of order $\mathcal{O}(\varepsilon k + k^2)$. By assumption, $\varepsilon << k$ and, therefore, it is a second order method.

In order to investigate the stability problems of (7.5b) we consider the scalar equation

$$y_{n+1} = (1 + 2k\varepsilon\tilde{b})y_{n-1} + 2kiby_n, \quad \tilde{b} = \tilde{b}_1 + i\tilde{b}_2, \quad b_1, b_2 \text{ real.} \tag{7.6}$$

The characteristic equation now becomes

$$\kappa^2 - 2kib\kappa - (1 + 2k\varepsilon\tilde{b}) = 0.$$

Therefore, for $1 - k^2b^2 >> |k\varepsilon\tilde{b}|$,

$$\begin{aligned}\kappa &= kib \pm \sqrt{1 - k^2b^2 + 2k\varepsilon\tilde{b}} \\ &= kib \pm \left(\sqrt{1-k^2b^2} + \frac{k\varepsilon\tilde{b}}{\sqrt{1-k^2b^2}} + \mathcal{O}(\varepsilon^2k^2)\right),\end{aligned}$$

i.e.,

$$\begin{aligned}|\kappa|^2 &= (kb \pm k\varepsilon\tilde{b}_2)^2 + 1 - k^2b^2 + 2k\varepsilon\tilde{b}_1 + \mathcal{O}(\varepsilon k^4 + \varepsilon^2k^2) \\ &= 1 \pm 2k^2\varepsilon b\tilde{b}_2 + 2k\varepsilon\tilde{b}_1 + \mathcal{O}(\varepsilon k^4 + \varepsilon^2k^2).\end{aligned}$$

Thus, neglecting terms of order $\mathcal{O}(\varepsilon k^4 + \varepsilon^2 k^2)$,

$$|\kappa| \le 1 \text{ if } \tilde{b}_1 \le -k|b\,b_2|.$$

(See Theorem 7.1.) If $\tilde{b}_1 = 0$, and $|b\,b_2| \neq 0$, then there is growth. However, $|\kappa|^2 = 1 + k^2\mathcal{O}(\varepsilon)$ and, therefore, the amplification is weak and does not destroy the calculations in time intervals of length $\mathcal{O}(1/\varepsilon)$. Similar results hold for systems.

8. General multistep methods. In this section we consider general approximations of the form

$$\left(I - \alpha_{-1}\frac{k}{\varepsilon}A_{n+1}\right)y_{n+1} = \sum_{j=0}^{r-1}\alpha_j y_{n-j} + k\sum_{j=0}^{q-1}\beta_j B_{n-j}y_{n-j}. \tag{8.1}$$

(For simplicity only, we assume that $F \equiv 0$.) We are interested in methods which are accurate of order p and we first look at the case when $B \equiv 0$. We determine the coefficients α_j such that

$$\begin{aligned}&u_{n+1} - k\alpha_{-1}\frac{du_{n+1}}{dt} - \sum_{j=0}^{r-1}\alpha_j u_{n-j} \\ &= c_{p+1}\frac{k^{p+1}}{(p+1)!}\frac{d^{p+1}u_{n+1}}{dt^{p+1}} + c_{p+2}\frac{k^{p+2}}{(p+2)!}\frac{d^{p+2}u_{n+1}}{dt^{p+2}} + \mathcal{O}(k^{p+3}),\end{aligned}$$

for all smooth functions. Using Taylor expansion, this leads to the linear equations

$$\sum_{j=0}^{r-1}\alpha_j = 1, \quad \sum_{j=0}^{r-1}(j+1)\alpha_j = \alpha_{-1}, \quad \sum_{j=0}^{r-1}(j+1)^l\alpha_j = 0, \quad l = 2, 3, \ldots, p. \tag{8.2}$$

Also,

$$c_\nu = -(-1)^\nu \sum_{j=0}^{r-1} (j+1)^\nu \alpha_j. \tag{8.3}$$

If we set $r = p$ we recover the BDF formulas (4.2).

Now assume that we have chosen α_j such that (8.2) holds. We will apply the method for the case that $A \equiv 0$. We want to choose the β_j such that the resulting method

$$y_{n+1} = \sum_{j=0}^{r-1} \alpha_j y_{n-j} + k \sum_{j=0}^{q-1} \beta_j B_{n-j} y_{n-j} \tag{8.4}$$

is accurate of order s with $p \le s \le q$. In this case $By = dy/dt$ and, therefore, we have to choose β_j such that, for all functions u,

$$u_{n+1} - \sum_{j=0}^{r-1} \alpha_j u_{n-j} - k \sum_{j=0}^{q-1} \beta_j \frac{du_{n-j}}{dt} = \mathcal{O}(k^{s+1}). \tag{8.5}$$

Using (8.2), we have,

$$\begin{aligned} u_{n+1} - &\sum_{j=0}^{r-1} \alpha_j u_{n-j} - k \sum_{j=0}^{q-1} \beta_j \frac{du_{n-j}}{dt} \\ = k\alpha_{-1} \frac{d_{n+1}}{dt} - &\sum_{l=p+1}^{s} \frac{(-1)^l k^l}{l!} \left(\sum_{j=0}^{r-1} (j+1)^l \alpha_j \right) \frac{d^l u_{n+1}}{dt^{l+1}} - k \left(\sum_{j=0}^{q-1} \beta_j \right) \frac{du_{n+1}}{dt} \\ - &\sum_{l=2}^{s} \frac{(-1)^l k^l}{(l-1)!} \sum_{j=0}^{q-1} (j+1)^{l-1} \beta_j \frac{d^l u_{n+1}}{dt^l} + \mathcal{O}(k^{s+1}). \end{aligned}$$

Therefore, (8.5) is satisfied if

$$\begin{aligned} &\sum_{j=0}^{q-1} \beta_j = \alpha_{-1}, \quad \sum_{j=0}^{q-1} (j+1)^{l-1} \beta_j = 0, \quad l = 2, \ldots, p, \\ &\frac{1}{l} \sum_{j=0}^{r-1} (j+1)^l \alpha_j = \sum_{j=0}^{q-1} (j+1)^{l-1} \beta_j, \quad l = p+1, \ldots, s. \end{aligned} \tag{8.6}$$

If the conditions (8.3), (8.6) are satisfied, then we obtain, for smooth functions,

$$L[u] =: \left(I - \alpha_{-1}\frac{k}{\varepsilon}A_{n+1}\right)u_{n+1} - \sum_{j=0}^{r-1}\alpha_j u_{n-j} - k\sum_{j=0}^{q-1}\beta_j B_{n-j}u_{n-j}$$

$$= k\alpha_{-1}\left(\frac{du_{n+1}}{dt} - \left(\frac{1}{\varepsilon}A_{n+1} + B_{n+1}\right)u_{n+1}\right) \tag{8.7}$$

$$+k^{p+1}R_{n+1}^{(p+1)} + k^{p+2}R_{n+1}^{(p+2)} + \mathcal{O}(k^{p+3}),$$

where

$$R^{(\nu)} = \frac{c_\nu}{\nu!}\frac{d^\nu u}{dt^\nu} + \frac{d_{\nu-1}}{(\nu-1)!}\frac{d^{\nu-1}(Bu)}{dt^{\nu-1}}. \tag{8.8}$$

Corresponding to (8.3),

$$d_\nu = -(-1)^\nu\sum_{j=0}^{q-1}(j+1)^\nu\beta_j \tag{8.9}$$

Thus, for slowly varying solutions, the error is at most $\mathcal{O}(k^{p+1})$.

As we have seen in the previous sections, the most interesting methods are obtained when $r = p = 2$ and $s = q = 2, 3, 4, 5, 6$. In this case

$$\alpha_{-1} = \frac{2}{3}, \quad \alpha_0 = \frac{4}{3}, \quad \alpha_1 = -\frac{1}{3}.$$

We have calculated the β_j in Table 8.1.

TABLE 8.1

q	β_0	β_1	β_2	β_3	β_4	β_5
2	1.3333	−0.6667				
3	1.7778	−1.5556	0.4444			
4	2.1667	−2.7222	1.6111	−0.3889		
5	2.5241	−4.1519	3.7556	−1.8185	0.3574	
6	2.8602	−5.8324	7.1167	−5.1796	2.0380	−0.3361

We will now discuss the stability of

$$\left(I - \frac{2}{3}\frac{k}{\varepsilon}A\right) u_{n+1} = \frac{4}{3}u_n - \frac{1}{3}u_{n-1} + kB\sum_{j=0}^{q-1}\beta_j u_{n-j}. \tag{8.10}$$

We assume that A, B are constant matrices.

If $A \equiv 0$, then the stability question is reduced to the usual scalar problem, i.e., the behavior of the roots of the characteristic equation

$$\kappa^q - \sum_{j=0}^{q-1}(\alpha_j + \lambda k\beta_j)\kappa^{q-j-1} = 0, \quad \alpha_j = 0 \text{ for } j > 1.$$

Since the basic characteristic equation ($\lambda k = 0$) is the same as for the BDF methods, we need only consider the perturbation

$$\kappa = e^{\lambda k} + k^{p+1}\tau.$$

For τ, $|\kappa|^2$ we obtain relations of type (5.12) and (5.13), respectively. For $q = 3, 4$, the methods are locally stable and (5.13a) holds. For $q = 2, 5, 6$, they are not locally stable. In Figure 8.1 we have calculated part of the stability region.

Now let $A \neq 0$. Without restriction we can assume that A has the blockdiagonal form (1.3). Then the first components ($\alpha_0 = 4/3$, $\alpha_1 = -1/3$, $\alpha_j = 0$ for $j > 1$) of the system satisfy

$$\left(I - \frac{2k}{3\varepsilon}A_{11}\right) y_{n+1}^I = \sum_{j=0}^{q-1}\left((\alpha_j I + k\beta_j B)y_{n-j}\right)^I. \tag{8.11}$$

As in the previous section, neglecting terms of order $\mathcal{O}(\varepsilon^2)$, the solution of (8.11) is

$$y_{n+1}^I = -\varepsilon\sum_{j=0}^{q-1}\beta_j A_{11}^{-1}B_{12}y_{n-j}^{II} + \tilde{y}^I, \tag{8.12a}$$

$$\left(I - \frac{2k}{3\varepsilon}A_{11}\right)\tilde{y}_{n+1}^I = \sum_{j=0}^{q-1}\alpha_j\tilde{y}_{n-j}^I. \tag{8.12b}$$

(8.12b) is the second order BDF method which is unconditionally stable. Neglecting terms of order $\mathcal{O}(\varepsilon)$, we obtain, for the other components,

$$\left(I - \frac{2k}{3}A_{22}\right)\tilde{y}_{n+1}^{II} = \sum_{j=0}^{q-1}(\alpha_j I + k\beta_j B_{22})y_{n-j}^{II}. \tag{8.13}$$

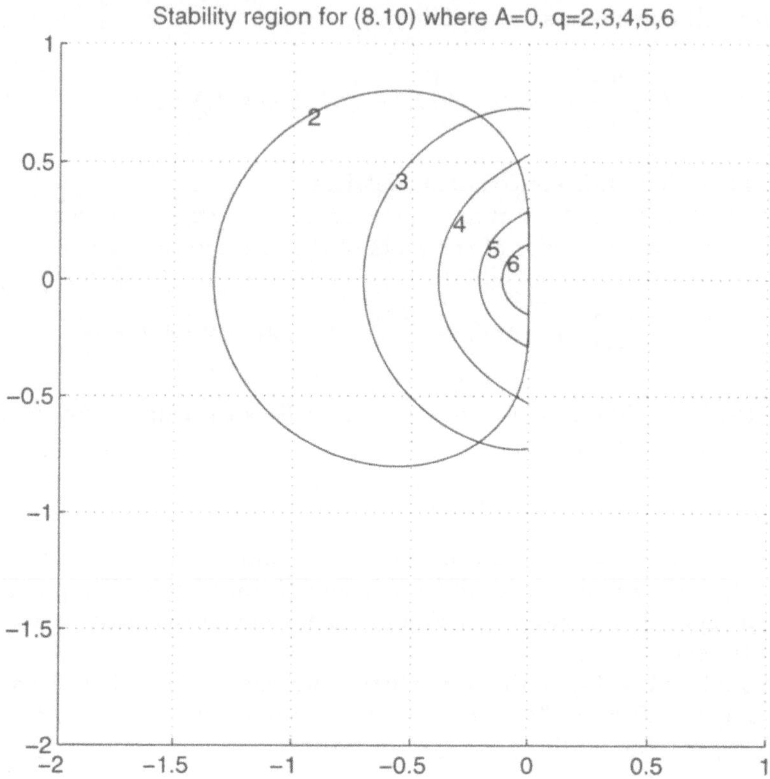

FIG. 8.1.

If $A_{22} = 0$, the above stability results apply. If $A_{22} \neq 0$ and A_{22} does not commute with B_{22}, we cannot reduce the stability question to a scalar equation. Instead, we must consider solutions of (8.13) which are of the form

$$y_n = \kappa^n y_0, \quad y_o \text{ vector}, \tag{8.14}$$

where

$$\kappa = e^{\lambda k} + (\lambda k)^3 \tau, \quad y_0 = u_0 + (\lambda k)^2 y_1,$$

and $u = e^{\lambda t} u_0$ is the corresponding solution of

$$u_t = (A_{22} + B_{22})u,$$

i.e.,

$$(A_{22} + B_{22})u_o = \lambda u_0. \tag{8.15}$$

Let

$$(\lambda k)^3 \tilde{R} = \left(\left(I - \frac{2k}{3} A_{22} \right) e^{\lambda q k} - \sum_{j=0}^{q-1} (\alpha_j I + k\beta_j B_{22})\, e^{\lambda(q-j-1)k} \right) u_0.$$

Introducing (8.14) into (8.13) and deleting the index 22, we obtain, using (8.2) and (8.6),

$$\begin{aligned}
0 = &\left(\left(I - \frac{2k}{3} A \right) (e^{\lambda k} + (\lambda k)^3 \tau)^q \right) \\
&- \left(\sum_{j=0}^{q-1} (\alpha_j I + k\beta_j B)(e^{\lambda k} + (\lambda k)^3 \tau)^{q-1-j} \right) (u_0 + (\lambda k)^2 y_1) \\
= &(\lambda k)^{p+1} \tilde{R} + (\lambda k)^3 \tau \left(q - \sum_{j=0}^{q-1} \alpha_j (q-j-1) \right. \\
&\left. + \lambda k \left(q(q-1) - \sum_{j=0}^{q-1} \alpha_j (q-(j+1))(q-1-(j+1)) \right) \right) u_0 \\
&- (\lambda k)^3 \tau \left(\frac{2}{3} qkA + kB \sum_{j=0}^{q-1} \beta_j (q-j-1) \right) u_0 \\
&+ (\lambda k)^2 \left(\lambda k q + \frac{1}{2} \lambda^2 q^2 k^2 - \lambda k \sum_{j=0}^{q-1} \alpha_j (q-j-1) \right. \\
&\left. - \frac{1}{2} \lambda^2 k^2 \sum_{j=0}^{q-1} \alpha_j (q-j-1)^2 \right) y_1 \\
&- (\lambda k)^2 k \left(\frac{2}{3} A (1 + \lambda q k) + B \sum_{j=0}^{q-1} \beta_j \left(1 + \lambda k (q-j-1) \right) \right) y_1 + \mathcal{O}(\lambda k)^5 \\
= &(\lambda k)^3 \tilde{R} + \frac{2}{3} (\lambda k)^3 \tau \left(1 + (2q-1)\lambda k \right) u_0 - \frac{2}{3} (\lambda k)^3 \tau q k (A+B) u_0 \\
&+ \frac{2}{3} (\lambda k)^3 (1 + \lambda q k) y_1 - \frac{2}{3} (\lambda k)^2 k (1 + \lambda q k)(A+B) y_1 + \mathcal{O}\left((\lambda k)^5 \right).
\end{aligned}$$

Therefore, (8.15) gives us,

$$\begin{aligned}
&\tau(1 - (q-1)\lambda k) u_0 + \frac{1}{\lambda}(1 + \lambda q k)(\lambda I - (A+B)) y_1 \\
&= -\frac{3}{2} \tilde{R} + \mathcal{O}(k^2).
\end{aligned} \tag{8.16}$$

We can express $\tilde{R}$ in terms of the truncation error. By (8.7), replacing

$\frac{1}{\varepsilon}A_{n+1}$ by A and B_{n-j} by B we obtain, using (8.15),

$$
\begin{aligned}
(\lambda k)^3\tilde{R} &= L[e^{\lambda t}u_0]e^{-\lambda(n+1-q)k} \\
&= \left(k^3R^{(3)}_{n+1} + k^4R^{(4)}_{n+1}\right)e^{-(n+1-q)k} + \mathcal{O}(k^5) \\
&= \frac{(\lambda k)^3}{3!}e^{\lambda qk}\left(c_3I + 3d_2\frac{1}{\lambda}B\right)u_0 + \frac{(\lambda k)^4}{4!}\left(c_4I + 4d_3\frac{1}{\lambda}B\right)u_0 \\
&= \frac{(\lambda k)^3}{3!}e^{\lambda qk}\left((c_3+3d_2)I - 3d_2\frac{1}{\lambda}A\right)u_0 \\
&\quad + \frac{(\lambda k)^4}{4!}\left((c_4+4d_3)I - 4d_3\frac{1}{\lambda}A\right)u_0 + \mathcal{O}(k^5).
\end{aligned}
\tag{8.17}
$$

We now make

Assumption 8.1. $A = -A^*$, $B = -B^*$ *are skew-Hermitean.*

Normalizing the eigenvector u_0 such that $|u_0| = 1$, we obtain, from (8.16) and (8.17),

$$
\tau = (\lambda k)^3\left(a_1 + a_2\left\langle u_0, \frac{1}{\lambda}Au_0\right\rangle\right) + (\lambda k)^4\left(b_1 + b_2\left\langle u_0, \frac{1}{\lambda}Au_0\right\rangle\right) + \mathcal{O}(k^5).
$$

Here the real constants a_j, b_j, $j = 1, 2$, can easily be calculated from the previous expressions. Therefore, $|\kappa|^2$ is of the form (5.13) and, corresponding to Theorem 5.1, we have

Theorem 8.1. *For sufficiently small* $k(|A_{22}| + |B_{22}|)$, *the methods*

$$
\left(I - \frac{2}{3}\frac{k}{\varepsilon}A\right)y_{n+1} = \sum_{j=0}^{q-1}(\alpha_jI + k\beta_jB)y_{n-j}
$$

are stabilized if we replace B *by* $B - \sigma\left(k(|A_{22}| + |B_{22}|)\right)^4 I$, *provided* σ *is sufficiently large.*

Since, for $q = 3$ and $A = 0$, the method is locally stable, we can use $\sigma = 0$ if $\langle\hat{u}_0, \frac{1}{\lambda}A_{22}\hat{u}_0\rangle$ is sufficiently small. As an example, we have considered the differential equation

$$
y' = (ig|c| + c)y, \quad c = c_1 + ic_2, \quad c_1, c_2 \text{ real}, \ c_1 \le 0,
$$

and approximated it by the combination of second order BDF with the third order explicit method

$$
\left(1 - \frac{2k}{3}ig|c|\right)y_{n+1} = \sum_{j=0}^{3}(\alpha_jI + k\beta_jc)y_{n-j}.
\tag{8.18}
$$

In Figure 8.2 we have calculated part of the stability region for different values of g.

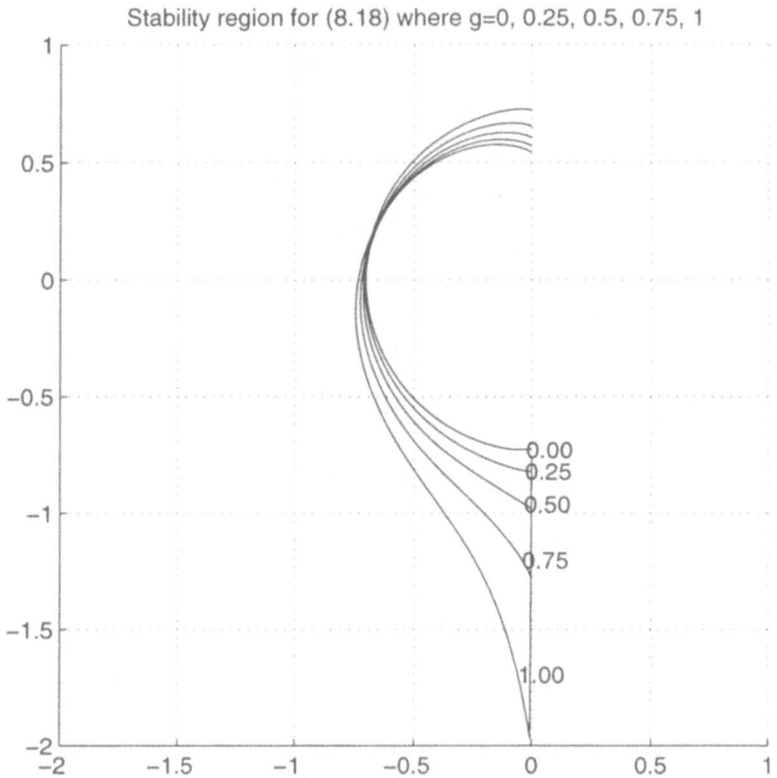

FIG. 8.2.

REFERENCES

[1] G. Browning and H.O. Kreiss, *Splitting methods for problems with different time scales*, Monthly Weather Rev., vol. **122**, (1994) pp. 2614–2622.

[2] B. Engquist, E. Haas and H.O. Kreiss, *Numerical solution of problems with different time scales I*, To appear.

[3] C.W. Gear, *Numerical Initial Value Problems in Ordinary Differential Equations*, Prentice-Hall, New Jersey, 1971.

[4] H.O. Kreiss, *Problems with different time scales for ordinary differential equations*, Siam J. Numer. Anal., vol. **6** No.6 (1979).

[5] H.O. Kreiss, *Problems with different time scales*, Acta Numerica, (1991), pp. 1–39.

[6] H.O. Kreiss and J. Lorenz, *On the existence of slow manifolds for problems with different time scales*, Phil. Trans. R. Soc. Lond., vol. **346**, (1994) pp. 159–171.

[7] H.O. Kreiss and L. Wu, *On the stability definition of difference approximations for the initial boundary value problem*, Appl. Num. Math., vol. **12** (1993) pp. 213–227.

[8] B. Lindberg, *On the smoothing and extrapolation for the trapezoidal rule*, Bit vol. **11** (1971) pp. 29–52.

MICROWAVE HEATING OF MATERIALS*

G.A. KRIEGSMANN†

1. Introduction. Although the classical fields of electromagnetic wave propagation and scattering theory have their roots in the last century, both have enjoyed vibrant lives over the past 50 odd years. Sophisticated asymptotic and numerical tools have been developed which give scientists and engineers the ability to solve problems of ever increasing complexity. This work has been spurred on by emerging and maturing technologies, primarily concerned with the transmission and retrieval of information. Mathematicians have made many significant and enduring contributions to these fields of applied science.

Other technologies have evolved over this span of time which depend upon the ability of electromagnetic waves to heat materials. They range from "low" to "high" technological applications. The former camp includes drying of wood, curing of paints, etc.[1]. The later embraces ceramic sintering and joining, microwave assisted chemical reactions and combustion, and the processing of polymers [2]. The ubiquitous cooking of foodstuffs by microwaves can be classified either way; low tech for boiling water and high tech for making a perfect basmati rice! These inventions and innovations have been made primarily by engineers and scientists.

The working engineering theories which describe electromagnetic heating are based upon heuristically averaged, linear equations [3]. These adequately explain some heating processes, such as microwave cooking of foodstuffs, but not others. These include such phenomenon as thermal runaway [4-6] and hot-spot formations [5-7] which have important ramifications in both biomedical and industrial applications. They are caused by the temperature dependencies of the electrical and thermal properties of the irradiated material which make the basic underlying mathematical description to be highly nonlinear. For example, the dependence of the effective electrical conductivity on temperature nonlinearly couples Maxwells equations to the heat equation while the presence of the electric and polarization fields in the source term of the heat equation couple it nonlinearly in the other direction.

There has recently been a considerable amount of work directed by mathematicians and scientists at the understanding of this nonlinear heating process. Most have dealt with one-dimensional models. The heating of a half-space has been examined in the low electrical conductivity limit

* This work was supported by the Air Force Office of Scientific Research Under Grant No. AFOSR F49620-94-1-0338, the Department of Energy Under Grant No. DE-FG02-94ER25196, and the National Science Foundation Under Grant No. DMS 9305828.

† Department of Mathematics, Center for Applied Mathematics and Statistics, New Jersey Institute of Technology, University Heights, Newark, NJ 07102.

[8,10], in the low thermal conductivity limit [11-13], and in models where these parameters were taken as powers of temperature, for a variety of thermal boundary conditions applied at the interface. Spatial structures corresponding to hot spots have been discovered in the second limit and response curves relating the final steady state temperature to the incident microwave power have been given for the first. These curves are monotonic and do not explain the phenomenon of thermal runaway.

The microwave heating of a finite slab has likewise recently received a considerable amount of study [14-16]. Unlike the infinite slab model, it has to transfer heat across both its faces to the environment and can not depend on "infinity" as a fixed temperature source. This fact coupled with the only assumption that the slab weakly loses thermal energy to its environment have given rise to an asymptotic theory which yields a multivalued response curve. This result explains the runaway phenomenon seen experimentally in ceramic sintering and also suggests a control process for alleviating its presence [4-6,16-17].

The microwave heating of a compact target has also been recently addressed [18] and the results are qualitatively the same as those of a finite slab. A somewhat terse review of this material is presented below. Although, the models that describe the heating of a slab or a compact target do capture a great deal of the correct physics, they do neglect one important feature. Targets are not irradiated by plane waves of a known strength E_0 and are not situated in an infinite, isotropic, and homogeneous environment; these are mathematical idealizations. The targets to be heated are usually placed in applicators or cavities where the electromagnetic energy is supplied by an incident mode. The amplitude of the electric field within the cavity is unknown and depends upon the geometry of the cavity and the interaction of the wave with the target. The determination of this amplitude is part of the problem and it adds another level of complexity. The effects of applicator physics upon a heating process have recently been analyzed for a slab in a TE_{103} cavity [19]. A very terse review of this material is presented below.

We shall now outline the remainder of this paper. Section II contains the mathematical formulation of a general heating problem and Section III describes a general numerical approach for attacking it. Section IV contains some results of an asymptotic theory which sheds a considerable amount of insight into the heating process. And finally Section V contains a description of some interesting cavity physics that affects the interaction of the electromagnetic waves with the sample.

2. Mathematical formulation. The formulation we present assumes a time harmonic electromagnetic fields and a time dependent temperature distribution. Although the governing equations do not admit exactly such a physical solution, the equations we present are in fact the leading order equations of an asymptotic theory. This theory is based upon the fact that the time required for heat to diffuse an electromagnetic wavelength is much larger than the period of a microwave. It yields the standard time harmonic version of Maxwell's equations for the electric and magnetic fields, and an averaged diffusion equation for the temperature distribution in which the electromagnetic source term has been integrated over a microwave period. A mathematical derivation of this theory using averaging can be found in Reference 18.

Within the framework of this theory, the electromagnetic field interacts with the material and increases its temperature through a combination of dipolar and ohmic heating. The temperature evolves according to the equation

$$\rho C_p \frac{\partial}{\partial t'} T = K \nabla^2 T + \frac{\sigma(T)}{2} |\mathbf{E}|^2, \qquad \mathbf{x}' \in \Omega \tag{2.1a}$$

where Ω is the region occupied by the material and K, ρ, and C_p are its thermal conductivity, density, and thermal capacity, respectively. Although these parameters depend upon the temperature, T, they are assumed constant in the following analysis. The reason for this assumption is the effective electrical conductivity, σ, of many important materials, such as ceramics, changes by one to two orders of magnitude when heated from room temperature to the required processing temperature. The changes in the other parameters (e.g., the thermal conductivity, K) are very small in comparison [18].

We also require that the temperature satisfies the surface heat balances

$$K \frac{\partial}{\partial n'} T + h(T - T_0) + se(T^4 - T_0^4) = 0, \qquad \mathbf{x}' \in \partial\Omega \tag{2.1b}$$

where h is the convective heat constant, s is the radiation heat constant, and e is the emissivity of the surface. Implicit in this model of surface heat transfer is the assumption that the sample sits in an infinite environment whose temperature is held fixed at T_0. Any model which attempts to describe the ability of the ceramic to heat its environment would have to take into account the finite size of the microwave applicator or cavity and these would in turn introduce new geometric effects. The inclusion of such realistic features and their qualitative effects would make an interesting and important research project. Finally, we assume that the sample is initially at the ambient temperature,

$$T(\mathbf{x}', 0) = T_0, \qquad \mathbf{x}' \in \Omega. \tag{2.1c}$$

Within the framework of the present theory, the time harmonic electromagnetic fields satisfy

$$\nabla X \mathbf{E} = i\omega\mu_0 \mathbf{H}, \qquad \mathbf{x}' \in D \tag{2.2a}$$

$$\nabla X \mathbf{H} = [-i\omega\epsilon + \sigma(T)]\mathbf{E}, \qquad \mathbf{x}' \in D \tag{2.2b}$$

where ω is the frequency of the microwave source and D is a region which includes Ω. It will be specified later. The effective conductivity σ can be extended to the region D by taking its value to be zero outside of the material. Similarly, the permittivity ϵ takes on its free-space value ϵ_0 outside of the material. We assume that the material is non-magnetic so that the permeability μ_0 is constant throughout D.

To complete the mathematical description of the problem we must specify boundary conditions for the electromagnetic fields. First, the tangential electric and magnetic fields must be continuous across the material-air interface. That is,

$$[\mathbf{E}X\hat{\mathbf{n}}]_{\partial\Omega} = [\mathbf{H}X\hat{\mathbf{n}}]_{\partial\Omega} = 0 \tag{2.2c}$$

where the $[\ ]_{\partial\Omega}$ denotes the jump in the quantity across the surface $\partial\Omega$. The remaining boundary conditions depend upon how the microwaves are delivered to the material sample. The simplest mathematical model would be to take the incident microwave to be a plane wave of known amplitude E_0 and the region $D = \mathbf{R}^3$. Then the electromagnetic fields would satisfy the far field boundary conditions

$$\left(\frac{\partial}{\partial r} + \frac{1}{r} - ik\right)\{\mathbf{E} - E_0 e^{ikx'}\hat{z}\} = O(r^{-3}), \quad r \to \infty \tag{2.2d}$$

$$\left(\frac{\partial}{\partial r} + \frac{1}{r} - ik\right)\{\mathbf{H} + E_0\sqrt{\epsilon_0/\mu_0}e^{ikx'}\hat{y}\} = O(r^{-3}), \quad r \to \infty \tag{2.2e}$$

where the terms in the $\{\}$ denote the scattered component of the fields. The exponential terms in the brackets denote the incident electric and magnetic fields, respectively, whose polarizations were chosen without loss of generality.

If the sample is situated in a cavity, then the boundary conditions and the problem are more involved. The region D will be the interior of the cavity and the tangential component of the electric field will be required to vanish on a portion of its boundary,∂D_1. The remaining portion of the boundary ∂D_2 connects the cavity to a source region or a waveguide carrying the microwave energy. If the tangential components of the electric and magnetic fields were known on ∂D_2, then (2.1) and (2.2a-2.2c) would give a well-posed problem whose solution would be qualitatively similar to the case where $D = \mathbf{R}^3$. However, the determination of these fields on ∂D_2 is part of the problem and their values depend intimately on what is happening in the cavity. We will not go into this matter here, but merely mention a result for a simple cavity later in this report.

So for the time being, the problem at hand is described by equations (2.1) and (2.2). The nonlinear character of this problem is apparent: The electric field impinges upon the material, penetrates it, and affects the temperature distribution through its presence in (2.1a). This in turn changes the effective electrical conductivity of the material, $\sigma(T)$, which affects the scattering process through its presence in (2.2b).

3. A numerical procedure. We shall now outline a straightforward numerical strategy for solving the coupled system (2a-2e). First, the heat equation is discretized in both space and time using, for example, finite differences. This is given symbolically by

$$A\mathbf{w}^{n+1} = \mathbf{F}(\mathbf{w}^n, \mathbf{e}^n) \tag{3.1a}$$

where $\mathbf{w}^n$ and $\mathbf{e}^n$ are the vectors that approximates T and $\mathbf{E}$ on the grid in Ω, respectively, and A is a linear operator. The superscript n denotes the time level corresponding to $t_n = n\delta t$. Next, a numerical method is chosen for the electromagnetic portion of the problem which is elliptic in character. Finite elements, for example, will produce a system given symbolically by

$$L(\mathbf{e}^n) + \sigma(\mathbf{w}^n)\mathbf{e}^n = \mathbf{f} \tag{3.1b}$$

where the forcing term $\mathbf{f}$ comes from the incident plane wave. It is important to note that the electric field and the temperature are evaluated at previous time steps in (3.1a) and (3.1b), respectively. This decouples the two systems. To see this, first set the initial temperature distribution $\mathbf{w}^0$ into (3.1b) with $n = 0$. The electric field is explicitly computed at time level t_0. This electric field is then inserted into (3.1a) with $n = 0$ and $\mathbf{w}^1$ is explicitly computed. This procedure is carried on until the temperature goes to a steady state or experiences a thermal runaway. This scheme has been used in similar and slightly more complex situations such as cavities and applicators [20-21]. No error or stability analyses have been performed.

Codes based upon the above scheme require extensive computer resources, especially when parameter studies are required to deduce trends and functional relationships. Moreover, samples are nearly insulated in many applications (to reduce nonuniform heating) and the heating process accordingly evolves on a long time scale fixed by the thermal and geometric properties of the material. This exacerbates the computational requirements of a numerical simulation.

4. An asymptotic theory. We have constructed an asymptotic approximation to the solution of (2) in the small Biot number limit, which often arises in many experiments [22-23]. This asymptotic theory gives a great deal of qualitative information about the heating process and forms a scientific basis by which large scale computations, such as the one described above, can be weighed and interpreted.

The Biot number for the heating process (2) is the dimensionless constant defined by $B = hL/K$ where L is a characteristic length of the sample, say the maximum diameter of Ω. This parameter is of the order of 10^{-3} in many experimental situations. There is a similar dimensionless parameter associated with radiative losses at the surface; it is $B_R = (seT_0^3)L/K$.

This is also a small number and we take it to be ordered as $B_R \sim B$. This insures that both radiative and convective heat loss mechanisms are built into the theory. Omitting the details, which can be found in References 16 and 18, the asymptotic analysis of (2) as $B \to 0$ yields

$$T = T_0[(1 + U_0(\tau)) + BU_1(\mathbf{x}, \tau) + O(B^2)] \tag{4.1}$$

where the leading order term satisfies the initial value problem

$$\frac{dU_0}{d\tau} = -g\{U_0 + \alpha[(U_0+1)^4 - 1]\} + pf(U_0) < |\mathsf{e}|^2 > \tag{4.2a}$$

$$U_0(0) = 0. \tag{4.2b}$$

The quantities in this equation are dimensionless and are defined by

$$\tau = t'/\theta, \quad \theta = L\rho C_P/h, \quad f = \sigma/\sigma_0, \quad \sigma_0 = \sigma(T_0), \quad p = \frac{1}{g}\left[\frac{\sigma_0 E_0^2}{2hT_0/L}\right]$$

$$\alpha = B_R/B, \quad \mathsf{e} = \frac{1}{E_0}\mathbf{E}, \quad g = SL/V$$

where S and V are the surface area and volume of Ω respectively. The time scale θ is a measure of how fast the material loses heat by convection, the parameter g is a geometric measure relating the surface area of the material to its volume, and p is a dimensionless power relating the power absorbed in the material to that lost on its surface. Finally, e is the dimensionless electric field scaled by the known amplitude of the incident wave E_0, and the average in (4.2a) is defined by

$$\langle |\mathsf{e}|^2 \rangle = \frac{L^3}{V} \int\int\int |\mathsf{e}|^2 \, dx\, dy\, dz \tag{4.3}$$

where the dimensionless spatial variables $\mathbf{x} \equiv \frac{1}{L}\mathbf{x}'$ and the integration is taken over the target.

It is very important to note that the leading order term in the temperature expansion U_0 is independent of $\mathbf{x}$. Thus, the electrical conductivity in the corresponding scaled version of Maxwell's equations, $f(U_0)$, depends only upon the time variable τ. This implies that the electric and magnetic fields depend only parametrically upon τ.

If the sample is a slab, a cylinder, or a sphere, then the electric field within the the material can be solved explicitly. The formula for e will be complicated and will depend explicitly upon $f(U_0)$. However, the average (4.3) can be computed and inserted into (4.2a) to yield an autonomous differential equation for U_0. To proceed further with the analysis of this simple differential equation a model for the effective electrical conductivity must be specified. The one we choose is $f(U_0) = e^{\chi U_0}$ which gives a good fit to experimentally measured conductivities for many low-loss ceramic materials [24-25]. The steady state solution of (4.2a) is shown in Figure 4.1 as an s-shaped curve relating U_0 to p. Recalling that $p \sim E_0^2$ this result states that there can be multiple temperature states for a give incident wave. A simple stability analysis shows that the upper and lower branches are stable while the middle is not. Moreover, we have shown that the upper branch is caused by the skin effect; the electrical conductivity becomes large at elevated temperatures and this shields the interior of the sample [14-16].

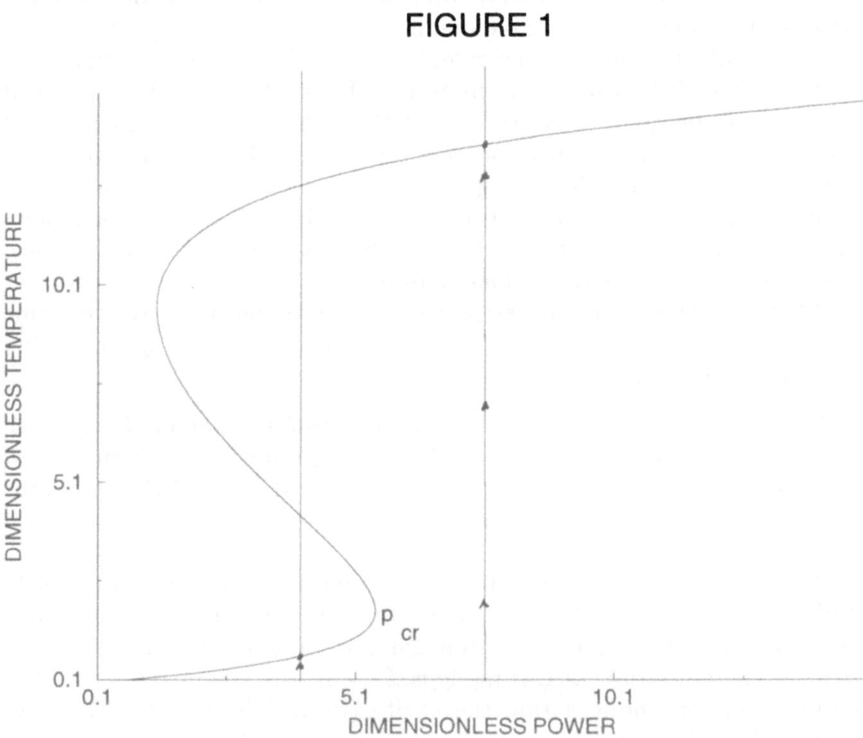

FIG. 4.1.

The vertical lines in Figure 4.1 illustrate the dynamics of the heating process in this limit. The state starts on the horizontal axis where $U_0(0) = 0$ corresponds to $T = T_0$. The system evolves to the lower branch as long as $p < p_{cr}$ and to the upper branch when $p > p_{cr}$. This discontinuous change in the final temperature as p_{cr} is crossed explains, in the context of this simplified theory, the phenomenon of thermal runaway which is seen in experiments [4-6].

Moreover, it is clear from this description that the entire middle branch and a portion of the upper branch are unreachable. If the required temperature for a process, such as sintering, lies on an inaccessible branch, then the process is doomed without further intervention. Several control theories have been presented to overcome this feature. In each the power p is changed at a rate dependent upon the temperature of the material and itself [17,26], i.e.,

$$\frac{dp}{d\tau} = G(p, U_0) \tag{4.4}$$

where G is the control function. A phase plane analysis of (4.2a-4.2b) and (4.4) for certain classes of G show that any temperature on the s-shaped curve can be obtained in a stable manner.

In general, the shape of the sample will preclude an exact expression for the electric field within the material. It can be argued on physical grounds that the qualitative features of the s-shaped response curve will be the same for any compact sample. However, this has not yet been established mathematically. The general numerical process of solving (4.2a-4.3) in conjunction with the scaled version of Maxwell's equations follows in a parallel fashion to that described in Section III. The only difference is that the heat equation is replaced by (4.2a-4.3), which is a significant simplification. Moreover, time steps are now chosen on the convective time scale θ rather than the diffusive time scale which tends to be much smaller in actual experiments [24-25].

5. Applicator effects. In this closing section we shall briefly describe a more realistic model of a microwave heating experiment and discuss a new qualitative feature of the heating process. The schematic of the apparatus is shown in Figure 5.1; it is a single mode microwave applicator composed of a rectangular waveguide and an iris.

The microwave source generates a complicated electromagnetic field which propagates down the wave guide. Since the dimensions of the guide allow only the lowest mode to propagate, we assume it is the only one present. The amplitude of this mode is E_0 which is assumed known. The amplitude of this mode within the cavity (formed by the iris, the wave guide, and the back wall) is unknown and its determination is part of the problem. This is the fundamental difference between the present problem and the one described in the previous sections.

The reason for putting the material in this cavity rather than in just a wave guide is one of efficiency, especially for low-loss materials. If the electrical conductivity of the material is very low at room temperature, then one way of increasing the power deposition rate $\sigma|\mathbf{E}|^2$ is to increase the amplitude of the electric field. By choosing a small aperture and a resonant cavity length, the electric field strength in an empty cavity can be quite large. However, as the sample heats its electrical conductivity increases, the cavity detunes, and the electric field strength diminishes to a state where efficient heating is no longer possible!

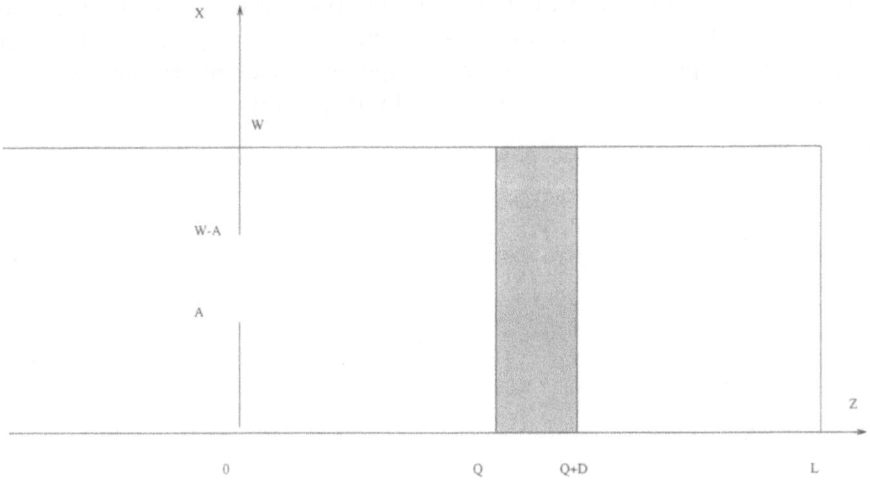

FIG. 5.1.

Finally, we have recently modeled this process in the small Biot number limit for a rectangular sample which completely fills the cross sectional area of the cavity [19]. It is beyond the scope of the present report to describe in much detail the analysis or results contained therein. However, we would like to point out the interesting fact that the leading order term in the expansion of the temperature satisfies a modified version of (4.2a-4.3). It is

$$\frac{dU_0}{d\tau} = -g\{U_0 + \alpha[(U_0+1)^4 - 1]\} + p\Gamma f(U_0) < |\mathbf{e}_1|^2 > \tag{5.1a}$$

$$U_0(0) = 0. \tag{5.1b}$$

$$\langle |\mathbf{e}_1|^2 \rangle = \frac{L^3}{V} \int\int\int |\mathbf{e}_1|^2 \, dx\, dy\, dz \tag{5.1c}$$

where $\mathbf{e}_1$ is the projection of the electric field along the lowest eigen mode and Γ is defined by

$$\Gamma = \frac{|1 + r_1|^2}{|1 - r_1\gamma_1|^2}. \tag{5.1d}$$

Here γ_1 is related to $\mathbf{e}_1$ evaluated at the interface of the slab and r_1 is the reflection coefficient for the iris in an infinite empty wave guide. Both of these terms depend upon the aperture opening. The determination of r_1 is a classical problem , so it is assumed know [27-28].

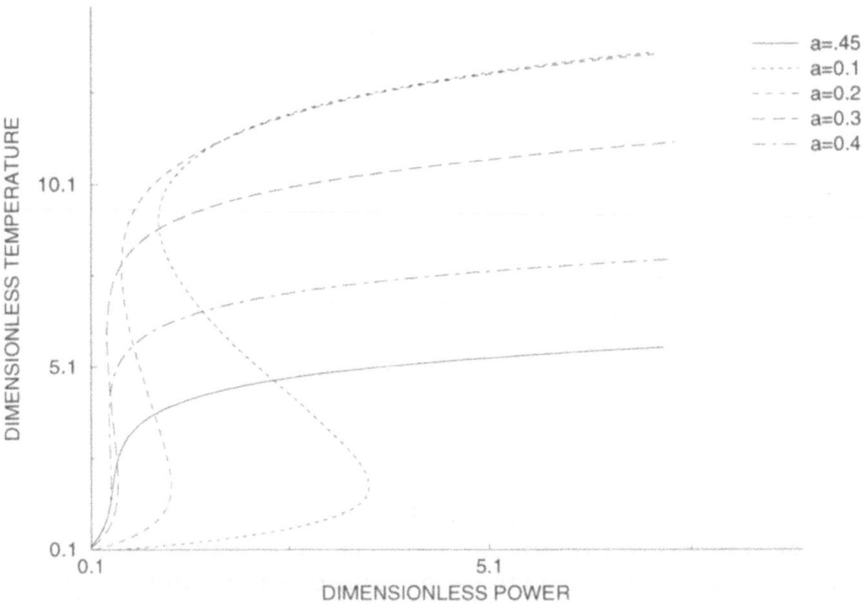

FIG. 5.2.

Several response curves corresponding to (4.4) for an exponential electrical conductivity are shown in Figure 5.2. There it is seen that increasing the iris height $a = A/W$ (i.e., decreasing the aperture) initially makes the heating process more efficient. That is, a smaller $p \sim E_0^2$ is required to obtain a fixed temperature. However, as the iris height increases further the upper branch begins to lower. Thus, a small aperture will initially increase the rate of heating, but the steady- state temperature will be lower than that achieved with no iris. This phenomenon is actually observed [29]. This effect, its consequences, and control are described fully in Reference 19.

REFERENCES

[1] D.A. Copson, *Microwave Heating*, A. VI Publishing Company, Inc., Westport, CT., 1975.

[2] M.F. Iskander, R.J. Lauf, and W.H. Sutton, eds., *Microwave Processing of Materials IV*, M.R.S. Symposium Proceedings, Vol. 347, 1994.

[3] A.C. Metaxas and R.J. Meredith, *Industrial Microwave Heating*, I.E.E. Power Engineering Series, Vol. 4, Peter Peregrimus Ltd., London, 1983.

[4] M.E. Brodwin and D.L. Johnson, *Microwave Sintering of Ceramics*, **MIT-S**, K-5 (1988), pp. 287–288.

[5] Y.L. Tian, D.L. Johnson, and M.E. Brodwin, *Ultrafine Structure of Al_2O_3 Produced by Microwave Sintering*, Proceedings of the First International Conference on Ceramic Powder Processing Science, Orlando Florida, Nov. 1987.

[6] A.J. Bertrand and J.C. Badot, *High Temperature Microwave Heating in Refractory Materials*, Journal of Microwave Power, 11 (1976), pp. 315–320.

[7] J.M. Hill and N.F. Smyth, *On the Mathematical Analysis of Hot-Spots Arising from Microwave Heating*, Math. Engng. Ind. 2 (1990), pp. 267–277.

[8] A.H. Pincombe and N.F. Smyth, *Microwave Heating of Materials with Low Conductivity*, Proc. Roc. Soc. 433 (1991), pp. 479–498.

[9] N.F. Smyth, *The Effect of Conductivity on Hot-spots*, J. Austral. Math. Soc. Ser. B, in press.

[10] G.A. Kriegsmann, D.Watters, and M.E. Brodwin, *Microwave Heating of a Ceramic Half-Space*, SIAM J. Appl. Math. 50 (1990), pp. 1088–1098.

[11] N.F. Smyth, *Microwave Heating of Bodies with Temperature Dependent Properties*, Wave Motion 12 (1990), pp. 171–186.

[12] T.R. Marchant and N.F. Smyth, *Microwave Heating of Materials with Non-Ohmic Conductance*, SIAM J. Appl. Math., submitted.

[13] J.M. Hill and A.H. Pincombe, *Some Similarity Temperature Profiles for the Microwave Heating of a Half-Space*, J. Austr. Math. Soc. (1992), pp. 290–320.

[14] G.A. Kriegsmann, *Microwave Heating of Ceramics*, in Ordinary and Partial Differential Equations 3, ed. B. Sleeman and R. Jarvis, Longman House 1991, pp. 45–56.

[15] G.A. Kriegsmann, *Microwave Heating of Ceramics: A Mathematical Theory*, in Microwaves: Theory and Applications in Materials Processing, ed., D.E. Clark, F.D. Gac, and W.H. Sutton, Ceramic Transactions 21, American Ceramic Society 1991, pp. 117–183.

[16] G.A. Kriegsmann, *Thermal Runaway in Microwave Heated Ceramics: A One-Dimensional Model*, Journal of Applied Physics, Vol. 71, No. 4 1992.

[17] G.A. Kriegsmann, *Thermal Runaway and Its Control in Microwave Heated Ceramics*, Microwave Processing of Materials III, eds. B.L. Beatty, W.H. Sutton, and M.F. Iskander, M.R.S. Symposium Proceedings, Vol. 269, 1992.

[18] G.A. Kriegsmann, *Microwave Heating of Dispersive Media*, SIAM Journal of Applied Mathematics, Vol. 53, No. 3, 1993.

[19] G.A. Kriegsmann, *Cavity Effects in Microwave Heating of Ceramics*, SIAM Journal of Applied Mathematics, submitted.

[20] R.L. Smith, M.F. Iskander, O. Andrade, and H. Kimrey, *Finite Difference Time Domain Simulations of Microwave Sintering in Multimode Cavities*, Microwave Processing of Materials III, eds. B.L. Beatty, W.H. Sutton, and M.F. Iskander, M.R.S. Symposium Proceedings, Vol. 269, 1992.

[21] M. Barmatz and H.W. Jackson, *Steady State Temperature Profiles in a Sphere Heated by Microwaves*, Microwave Processing of Materials III, eds. B.L. Beatty, W.H. Sutton, and M.F. Iskander, M.R.S. Symposium Proceedings, Vol. 269, 1992.

[22] W.D. Kingery, H.K. Bowen, and D.R. Uhlman, *Introduction to Ceramics*, Wiley, N.Y., 1976.

[23] F.P. Incorpera and D.P. Dewitt, *Introduction to Heat Transfer*, Wiley, NY., 1985.

[24] D.G. Watters, *An Advanced Study of Microwave Sintering*, Ph.D. Dissertation, Northwestern University, 1988.

[25] W.B. Westphal, *Dielectric constant and loss measurements on high-temperature materials*, Laboratory for Insulation Research Technical Report, MIT, October 1963.

[26] G.O. Beale and F.J. Arteaga, *Automatic Control to Prevent Thermal Runaway During Microwave Joining of Ceramics*, Microwave Processing of Materials III, eds. B.L. Beatty, W.H. Sutton, and M.F. Iskander, M.R.S. Symposium Proceedings, Vol. 269, 1992.

[27] D.S. Jones, *The Theory of Electromagnetics*, Pergamon Press Ltd., Oxford, 1964.

[28] L. Lewin, *Theory of Waveguides*, Halsted Press, John Wiley and Sons, New York, 1975.

[29] S.S. Sa'adaldin, W.M. Black, I. Ahmad, and R. Silberglitt, *Coupling with an Adjustable Compound Iris in a Single Mode Applicator*, Microwave Processing of Materials III, eds. B.L. Beatty, W.H. Sutton, and M.F. Iskander, M.R.S. Symposium Proceedings, Vol. 269, 1992.

CONTROL REGION APPROXIMATION FOR ELECTROMAGNETIC SCATTERING COMPUTATIONS

BRIAN J. McCARTIN*

Abstract. The Control Region Approximation is a generalized finite-difference procedure that accomodates completely general geometries and materials. It involves the discretization of conservation form equations on Dirichlet/Delaunay tessellations. This ensures the satisfaction of appropriate jump conditions across interfaces and permits a straightforward application of relevant boundary conditions. After presenting the basics of this discretization strategy, this paper details its application to electromagnetic scattering computations. Specific application is made to

- two-dimensional frequency-domain simulation,
- two-dimensional time-domain simulation,
- treatment of two-dimensional anisotropic media,
- three-dimensional frequency-domain simulation,
- two-dimensional periodic structures.

Finally, related ongoing work is described.

Key words. Dirichlet/Delaunay tessellations, control region approximation, finite-difference methods, electromagnetic scattering computations

AMS(MOS) subject classifications. 65C20, 65N05, 78A45, 78A55

1. Introduction. The scattering of electromagnetic (EM) waves presents myriad computational challenges which must be overcome prior to performing numerical simulations. First of all, the geometries about and within which scattering takes place are usually quite complicated and varied. Typical configurations include complete aircraft and human beings. Additionally, the material constitution of the scatterers can be complex, encompassing dielectrics, conductors, and magnetic materials. Furthermore, the media involved might be anisotropic and/or dispersive.

In addition to the above physical complications, there are mathematical/computational difficulties as well. The boundary value problem for electromagnetic scattering is not complete unless the asymptotic behavior of the scattered field is specified. Computationally, this translates into the need to enforce an absorbing boundary condition at a finite distance from the scatterer. Also, the frequencies of interest might entail a scatterer very large relative to free-space wavelength thus requiring a very fine mesh for adequate resolution. Moreover, the mesh width might also need to vary greatly over the solution domain due to either thin coatings on the surface of the scatterer or the presence of thin structures within the scatterer. Lastly, multiple incidence angles might be of interest in which case the computer resources required will increase accordingly.

In this paper, we present the Control Region Approximation [1,2,3,4,5]

* Science and Mathematics Department, GMI Engineering and Management Institute, 1700 West Third Avenue, Flint, MI 48504-4898, *bmccarti@nova.gmi.edu*

which is a physically-based discretization method addressing the above difficulties in a natural and efficient manner. This method has found wide application in engineering simulation. Caspar [6] has applied it to transonic aerodynamics while McCartin [7] has applied it to semiconductor device simulation. In the ensuing sections, we review its application to computational electromagnetics (CEM) [8] .

We begin with two-dimensional frequency-domain simulation. Here, we detail our basic spatial discretization technique encompassing general inhomogeneous materials. We also consider the enforcement of impedance and radiation boundary conditions. We then compare our computed results to experimental data for a spar-shell airfoil.

We continue with two-dimensional time-domain simulation. Here, we present a formulation that makes essential use of the duality of the Dirichlet/Delaunay tessellations to discretize Maxwell's equations. We then make a comparison of our computed results with an eigenfunction expansion for a cylinder coated with lossy magnetic material.

We next proceed to a treatment of two-dimensional anisotropic media. Again, duality allows us to discretize the problem in a physically satisfying manner. We then compare our computed results for an anisotropic rod to a published solution.

We then address three-dimensional frequency-domain simulation. We first recast Faraday's and Ampere's laws in circuital form. Once again, we invoke duality in order to provide a natural discretization thereof. We then compare computed results for a PEC cube with published time-domain results.

Finally, we consider two-dimensional periodic structures. Here, we reduce the problem to the unit cell of the grating by invoking Floquet's theorem. We further reduce the computational domain by applying an absorbing boundary condition to the propagating modes a short distance away from the grating. Then we study a "webbed" configuration which is a model of a jet engine cooling passage.

We conclude by outlining ongoing work which is focused on extending and applying this powerful simulation methodology.

2. Control region approximation. The first stage in the Control Region Approximation is the tessellation of the solution domain by Dirichlet regions associated with a pre-defined yet arbitrary distribution of grid points. Denoting a generic grid point by P_i, we define its Dirichlet region as

$$D_i = \{P : \| P - P_i \| < \| P - P_j \|, \forall j \neq i\}. \tag{2.1}$$

This is seen to be the convex polygon formed by the intersection of the half-spaces defined by the perpendicular bisectors of the the straight line segments connecting P_i to P_j. It is the natural control region to associate with P_i since it contains those and only those points which are closer to P_i

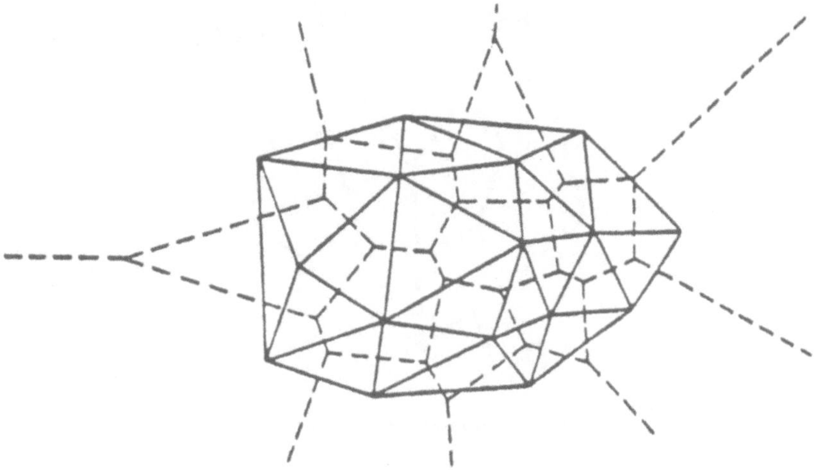

FIG. 2.1. *Dirichlet/Delaunay Tessellations*

than to any other grid point.

If we construct the Dirichlet region surrounding each of the grid points, we obtain the Dirichlet tessellation which is shown dashed in Figure 2.1. There we have also connected by solid lines neighboring grid points which share an edge of their respective Dirichlet regions. This construction tessellates the convex hull of the grid points by so-called Delaunay triangles. The union of these triangles is referred to as the Delaunay tessellation. It is essential to note that these two tessellations are dual to one another in the sense that corresponding edges of each are orthogonal.

With reference to Figure 2.2, we will exploit this duality in order to approximate the cross-derivative u_{xy} at the point P^*. We begin by integrating over the Dirichlet region, rewriting as a divergence, and applying the divergence theorem, thereby producing

$$\int\int_D u_{xy}\, dA = \int\int_D \nabla \cdot \begin{bmatrix} 0 \\ u_x \end{bmatrix} dA = \oint_{\partial D} \begin{bmatrix} 0 \\ u_x \end{bmatrix} \cdot \vec{\nu}\, d\sigma \tag{2.2}$$

where (ν, σ) are normal and tangential coordinates, respectively, around the periphery of D. Transforming to this coordinate system and making a piecewise-constant approximation along the edges of D yields

$$u^*_{xy} \approx \frac{1}{A} \sum_m \nu_2^{(m)} \left(\nu_1^{(m)} u_\nu^{(m)} - \nu_2^{(m)} u_\sigma^{(m)}\right) \Delta\sigma^{(m)} \tag{2.3}$$

where the index m ranges over the sides of D.

In the above expression, the flux term, u_ν, may be approximated by a central difference, while the shear term, u_σ, may be approximated by a central difference after (e.g. bilinear) interpolation to the vertices of

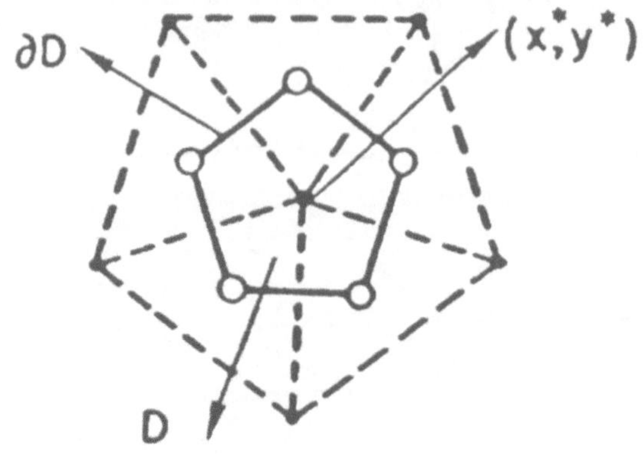

FIG. 2.2. *Numerical Differentiation/Integration*

D. Alternatively, the grid points may be modified to include also these vertices and the above procedure supplemented by integration over the Delaunay triangles. Note that in many physical problems, including those considered in later sections of this paper, the shear terms are absent. In addition, the numerical differentiation scheme just described is applicable to the approximation of any derivative.

Furthermore, we may perform numerical integration over arbitrary domains in a similar fashion. After having decomposed the domain of integration into Dirichlet regions, we may then expand the integrand in a Taylor series within each of them yielding the approximation

$$\begin{aligned}(2.4)\quad \iint_D F(x,y)\,dA \approx F^* \cdot \iint_D 1\,dA + F_x^* \cdot \iint_D (x-x^*)\,dA \\ +F_y^* \cdot \iint_D (y-y^*)\,dA + \frac{1}{2}F_{xx}^* \cdot \iint_D (x-x^*)^2\,dA \\ +F_{xy}^* \cdot \iint_D (x-x^*)(y-y^*)\,dA + \frac{1}{2}F_{yy}^* \cdot \iint_D (y-y^*)^2\,dA.\end{aligned}$$

In the above, the integrals are evaluated analytically while the derivatives are estimated using the previously described procedure.

With reference to Figure 2.3, we now compare this discretization strategy to the Finite Element Method (FEM) using bilinear elements. If we discretize the Poisson equation

$$(2.5)\qquad \Delta u = f$$

at P_0, the Control Region Approximation is

$$(2.6)\qquad (u_1 + u_2 + u_3 + u_4 - 4u_0) = h^2 f_0$$

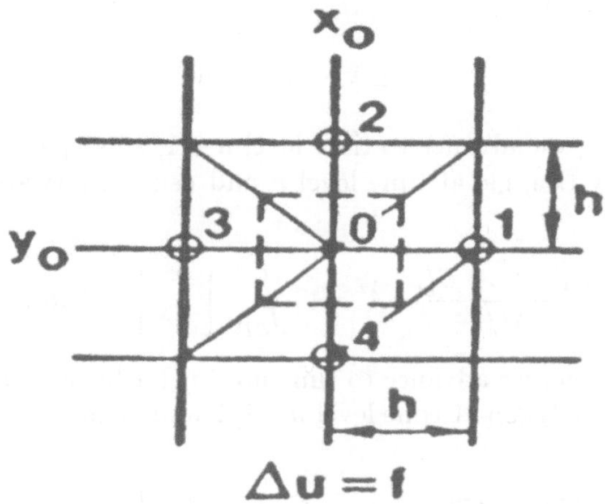

FIG. 2.3. *Comparison to Finite Element Approximation*

while the FEM approximation is

$$(u_1 + u_2 + u_3 + u_4 - 4u_0) = \frac{N}{6} h^2 f_0 \tag{2.7}$$

where N is the number of triangles with P_0 as vertex. Thus, the Control Region Approximation provides a local flux balance while the FEM provides only a global flux balance.

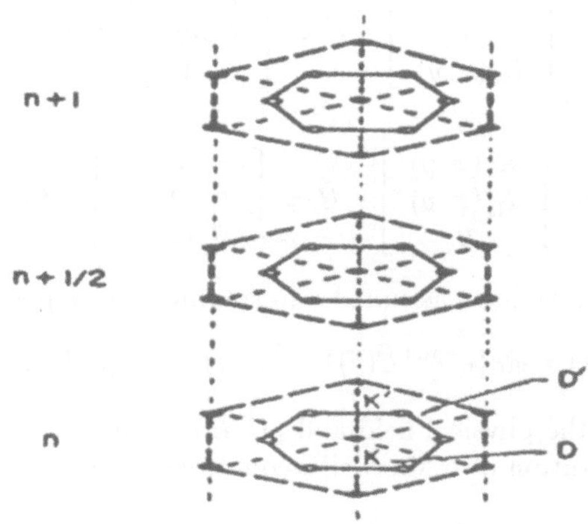

FIG. 2.4. *Two-Step Lax-Wendroff Method*

Temporal discretization may also be effected by this method. Figure 2.4 displays the spatio-temporal grid configuration. We illustrate the procedure with the two-step Lax-Wendroff method applied to the conservation law

$$U_t + F_x + G_y = 0. \tag{2.8}$$

In the first stage, we advance to time level $n + 1/2$ by spatially integrating over a Delaunay triangle at time level n and using a forward difference in time

$$\frac{U_{k'}^{n+1/2} - U_{k'}^{n}}{\Delta t/2} \cdot A' = -\oint_{\partial D'} \begin{bmatrix} F \\ G \end{bmatrix}^n \cdot \vec{\nu}\, d\sigma. \tag{2.9}$$

In the second stage, we advance to time level $n+1$ by spatially integrating over a Dirichlet polygon at time level $n+1/2$ and using a central difference in time

$$\frac{U_k^{n+1} - U_k^n}{\Delta t} \cdot A = -\oint_{\partial D} \begin{bmatrix} F \\ G \end{bmatrix}^{n+1/2} \cdot \vec{\nu}\, d\sigma. \tag{2.10}$$

In this fashion, we are able to extend this second order accurate time-marching scheme to unstructured meshes.

3. Two-Dimensional Frequency-Domain Simulation. Consider the scattering of a two-dimensional electromagnetic wave by a cylindrical obstacle of arbitrary cross-section (Figure 3.1). The general solution of such a problem can be obtained as the superposition of Tranverse Magnetic (TM) and Transverse Electric (TE) components [9,10] where

$$\vec{E} = \begin{bmatrix} 0 \\ 0 \\ E_z(x,y) \end{bmatrix};\ \vec{H} = \begin{bmatrix} H_x(x,y) \\ H_y(x,y) \\ 0 \end{bmatrix}\ (TM), \tag{3.1}$$

$$\vec{E} = \begin{bmatrix} E_x(x,y) \\ E_y(x,y) \\ 0 \end{bmatrix};\ \vec{H} = \begin{bmatrix} 0 \\ 0 \\ H_z(x,y) \end{bmatrix}\ (TE). \tag{3.2}$$

In the above, we have assumed a harmonic time dependence

$$\tilde{E}(\vec{r},t) = \Re e\{e^{-j\omega t}\vec{E}(\vec{r})\},\ \tilde{H}(\vec{r},t) = \Re e\{e^{-j\omega t}\vec{H}(\vec{r})\} \tag{3.3}$$

where $(\tilde{E}, \tilde{H})$ is the physical field and $(\vec{E}, \vec{H})$ is the phasor field.

Upon substitution into Maxwell's equations, we obtain the Helmholtz-type equation

$$\nabla_t \cdot (a\nabla_t u) + bu = 0 \tag{3.4}$$

with ∇_t the transverse gradient and

$$u = E_z,\ a = \frac{1}{\mu},\ b = \omega^2\epsilon\ (TM);\ u = H_z,\ a = \frac{1}{\epsilon},\ b = \omega^2\mu\ (TE), \tag{3.5}$$

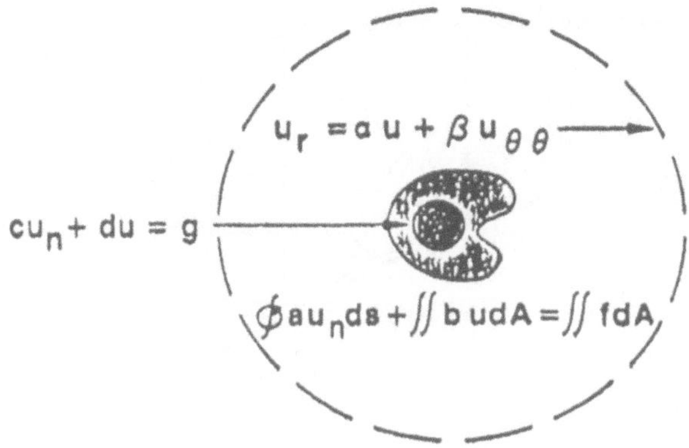

FIG. 3.1. *Scattering Problem*

where ϵ and μ are the complex permittivity and permeability, respectively, each of which may be arbitrary functions of position. Thus, the problem is reduced to the solution of scalar Helmholtz-type equations.

These equations must be supplemented by appropriate boundary conditions at the surface of any conductors. In particular, we may incorporate the impedance boundary condition

$$\vec{E} - (\vec{E} \cdot \hat{n})\hat{n} = \eta_s \hat{n} \times \vec{H} \tag{3.6}$$

where η_s is the surface impedance. Under our previous assumptions on the field, this condition reduces to

$$\frac{\partial u}{\partial n} - \gamma u = 0 \tag{3.7}$$

with

$$\gamma = -j\frac{\omega\mu}{\eta_s} \ (TM); \ \gamma = +j\omega\epsilon\eta_s \ (TE) \tag{3.8}$$

and $\hat{n}$ is the unit normal to the cross-section. Note that in the case of a perfect electrical conductor (PEC: $\eta_s = 0$), this reduces to

$$u = 0 \ (TM); \ u_n = 0 \ (TE). \tag{3.9}$$

As it stands, this formulation in not sufficient to guarantee a unique solution. An additional condition is needed to insure that the scattered field is composed of outgoing waves only. Incoming scattered waves are excluded by the Sommerfeld radiation condition

$$\frac{\partial u_S}{\partial r} - jku_S = O(r^{-1/2}); \ k^2 = \omega^2\mu\epsilon \tag{3.10}$$

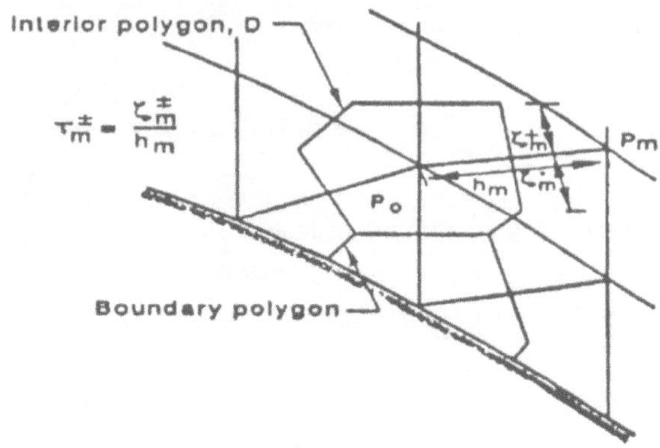

FIG. 3.2. *Discretization Procedure*

where the total field has been decomposed into the sum of an incident field, u_I, and a scattered field, u_S,

$$u = u_I + u_S. \tag{3.11}$$

This condition is approximated by an absorbing boundary condition [11,12] on either a circular or elliptical contour surrounding the scatterer.

With reference to Figure 3.1, the formulation reduces to the following exterior boundary value problem (BVP) for the scattered field:

$$\nabla \cdot (a\nabla u_S) + bu_S = F := -[\nabla \cdot (a\nabla u_I) + bu_I] \text{ in } \Omega, \tag{3.12}$$

$$\nabla u_S \cdot \hat{n} - \gamma u_S = -[\nabla u_I \cdot \hat{n} - \gamma u_I] \text{ on } C_i, \tag{3.13}$$

$$\frac{\partial u_S}{\partial r} + \alpha u_S + \beta \frac{\partial^2 u_S}{\partial \theta^2} = 0 \text{ on } C_0. \tag{3.14}$$

In the above, a, b, and γ are (possibly discontinuous) spatially varying complex functions while α and β are complex constants. We next discretize this BVP via the Control Region Approximation.

We first reformulate the problem in integral (conservation) form by integrating over a control region, D, and applying the divergence theorem, resulting in (see Figure 3.2)

$$\oint_{\partial D} a\frac{\partial u_S}{\partial \nu}\, d\sigma + \int\int_D bu_S\, dA = \int\int_D F\, dA. \tag{3.15}$$

The distinct advantage of this reformulation is that is still applies when the coefficients are discontinuous as they typically are in applications involving layered media.

After performing the numerical integrations described in the previous section, we arrive at the following discrete equation at the point P_0

$$\sum_m (\tau_m^- a_m^- + \tau_m^+ a_m^+)(u_m - u_0) + \sum_m b_{m,0}A_{m,0}u_0 = \sum_m F_{m,0}A_{m,0}. \tag{3.16}$$

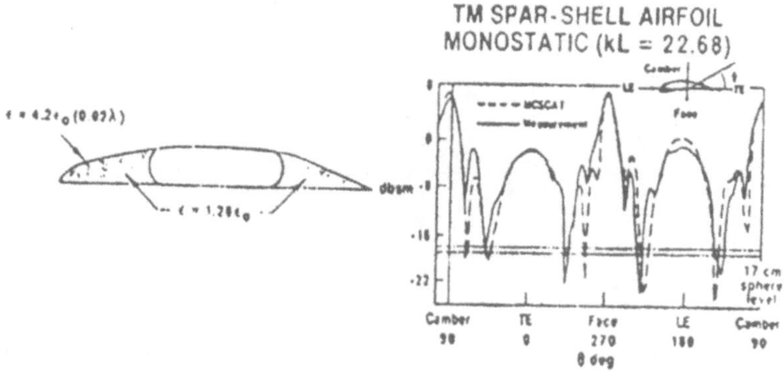

FIG. 3.3. *Spar-Shell Airfoil*

In the above, F must include appropriate δ-functions along the triangle edges due to discontinuities in a. Moreover, since both surface and absorbing boundary conditions involve either the field or its normal flux, they are readily accomodated. It is important to observe that this discretization produces sparse equations which can thus be solved very efficiently.

The radar cross-section (RCS), $\sigma(\theta)$, is defined as the scattered power per unit length in a fixed direction, θ, normalized by that of the incident field. For either TM or TE waves, this can be written as

$$\sigma(\theta) = \lim_{r \to \infty} 2\pi r \mid u_S \mid^2 \tag{3.17}$$

where we have assumed that $\mid u_I \mid = 1$ as it is, for example, in the case of a uniform plane wave.

The computed TM radar cross-section of a spar-shell airfoil is compared with measured data in Figure 3.3. The outer radius of the boundary is located at $1.5a$, where a is the semi-chord width. The outer shell has $\epsilon = 4.2\epsilon_0$ and its thickness is 0.02λ. The spar is metal and the dielectric fill is lossless with $\epsilon = 1.28\epsilon_0$. The computed RCS is used to estimate the RCS of the 19 in. long experimental model by neglecting the scattering from the end caps. The RCS of the finite cylinder is approximated by

$$\sigma_{3D} \cong 8(h^2/\lambda) \cdot \sigma_{2D} \tag{3.18}$$

where $2h =$ cylinder length.

Results are shown for an airfoil with 3.5λ chord width. The differences between the computed and measured returns is likely ascribable to leading- and trailing-edge details of the airfoil which were not modelled in the analysis. Most of the features of the scattering pattern are quite accurately reproduced by this finite-difference simulation. The overall agreement is quite favorable and is indicative of the useful engineering information obtainable from such two-dimensional analyses.

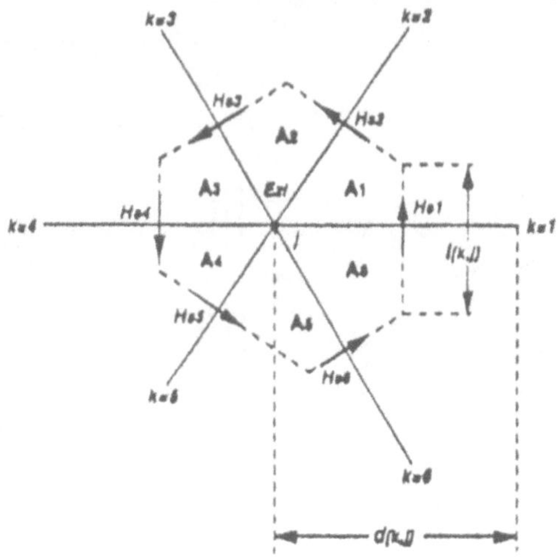

FIG. 4.1. *Time-Domain Dual Formulation*

4. Two-dimensional time-domain simulation. There are certain inherent benefits of performing simulations in the time-domain. Firstly, the impulse response can be so calculated and hence all target information can be captured in compact form. Secondly, the time-history of the scattering response more readily permits identification of scattering mechanisms. Finally, time-domain simulation is readily adapted to parallel computation for large three-dimensional problems. However, for a fixed frequency, frequency-domain simulations usually provide more accurate prediction of RCS. Moreover, if the materials of interest are dispersive then frequency-domain simulation is more easily performed. Thus, time-domain and frequency-domain simulations are properly viewed as complementary engineering prediction tools.

Figure 4.1 illustrates the time-domain formulation for electric field polarization (TM). In this case, with (n, s) normal/tangential coordinates around a control region, Maxwell's equations are

$$\oint_{\partial D} \eta_0 H_s \, ds = \int\int_D (\eta_0 \sigma_e E_z + \epsilon_r \frac{\partial E_z}{c\partial t}) \, dA \tag{4.1}$$

$$\frac{\partial E_z}{\partial n} = \sigma_m H_s + \mu_r \eta_0 \frac{\partial H_s}{c\partial t} \tag{4.2}$$

where ϵ_r, μ_r, σ_e, σ_m, η_0, and c are the relative permittivity, relative permeability, electric conductivity, magnetic conductivity, free-space impedance, and speed of light in free-space, respectively.

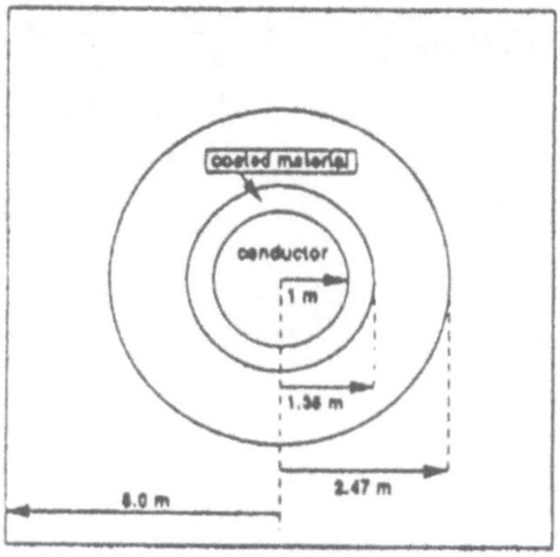

FIG. 4.2. *Coated Circular Cylinder*

Suitably discretized, these become

$$\sum_k \left(\frac{\eta_0 \sigma_e A_k}{2} + \frac{\epsilon_r A_k}{c\Delta t}\right) E^m_{z[j]} = -\sum_k \left(\frac{\eta_0 \sigma_e A_k}{2} - \frac{\epsilon_r A_k}{c\Delta t}\right) E^{m-1}_{z[j]} + \sum_k \eta_0 H^{m-1/2}_{s[k,j]} l_{[k,j]} \tag{4.3}$$

$$\left(\frac{\sigma_m}{2} + \frac{\mu_r \eta_0}{c\Delta t}\right) H^{m+1/2}_{s[k,j]} = -\left(\frac{\sigma_m}{2} - \frac{\mu_r \eta_0}{c\Delta t}\right) H^{m-1/2}_{s[k,j]} + \frac{1}{d_{[k,j]}}\left(E^m_{z[k]} - E^m_{z[j]}\right) \tag{4.4}$$

where E_z is defined at the center of a control region and H_s is defined along its edges. In the above, E_z is defined at integer time levels, $t = m\Delta t$, while H_s is defined at half-integer time levels, $t = (m + 1/2)\Delta t$.

For magnetic field polarization (TE), we may reverse the roles of E and H to obtain Maxwell's equations as

$$\oint_{\partial D} E_s \, ds = -\int\int_D \left(\sigma_m H_z + \mu_r \eta_0 \frac{\partial H_z}{c\partial t}\right) dA \tag{4.5}$$

$$\eta_0 \frac{\partial H_z}{\partial n} = -\eta_0 \sigma_e E_s - \epsilon_r \frac{\partial E_s}{c\partial t} \tag{4.6}$$

which upon discretization become

$$\sum_k \left(\frac{\sigma_m A_k}{2} + \frac{\mu_r \eta_0 A_k}{c\Delta t}\right) H^m_{z[j]} = -\sum_k \left(\frac{\sigma_m A_k}{2} - \frac{\mu_r \eta_0 A_k}{c\Delta t}\right) H^{m-1}_{z[j]} - \sum_k E^{m-1/2}_{s[k,j]} l_{[k,j]} \tag{4.7}$$

$$
(4.8) \qquad (\frac{\sigma_e \eta_0}{2} + \frac{\epsilon_r}{c\Delta t}) E_{s[k,j]}^{m+1/2} = -(\frac{\sigma_e \eta_0}{2} - \frac{\epsilon_r}{c\Delta t}) E_{s[k,j]}^{m-1/2} - \frac{\eta_0}{d_{[k,j]}} (H_{z[k]}^m - H_{z[j]}^m).
$$

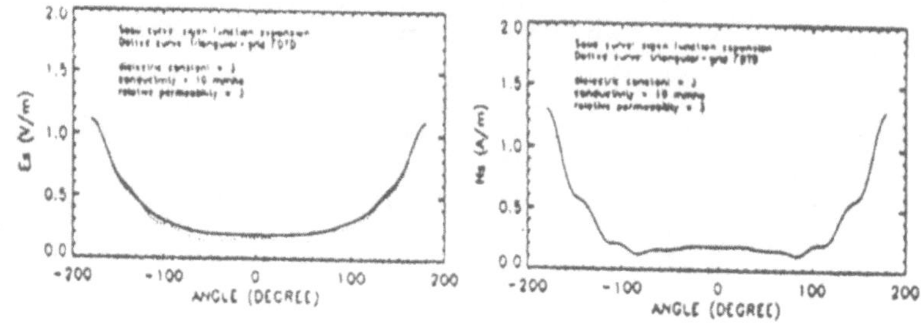

FIG. 4.3. *Coated Cylinder - Magnetic Material*

The stability criterion for this scheme is [13]

$$
(4.9) \qquad \Delta t \leq \min_j \left(\frac{1}{c} \sqrt{\frac{2A_j}{\sum_k \frac{l_{[k,j]}}{d_{[k,j]}}}} \right).
$$

This same reference discusses the absorbing boundary condition used.

We now analyze the scattering by a magnetically coated cylinder with loss as shown in Figure 4.2. The results are presented as a function of frequency which are extracted from the impulse response. The parameters of the coating are $\epsilon_r = 3$, $\sigma_e = 10$mmho/m, and $\mu_r = 3$. Figure 4.3 displays the fields for both polarizations at 2.47m and 150MHz. The dotted lines represent our numerical results and the solid lines represent an eigenfunction expansion. Any deviations are likely due to the fact that only seven grid points per wavelength were used within the coating.

5. Two-dimensional anisotropic media. With the field assignments displayed in Figure 5.1, we write Maxwell's equations in the circuital form

$$
(5.1) \qquad \oint_{\partial D} \vec{E} \cdot d\vec{l} = -\int\int_D j\omega \vec{B} \cdot d\vec{A}
$$

$$
(5.2) \qquad \oint_{\partial D'} \vec{H} \cdot d\vec{l} = \int\int_{D'} (j\omega \vec{D} + \vec{J}) \cdot d\vec{A}.
$$

We will treat the simplest type of nontrivial anisotropy

$$
(5.3) \qquad \bar{\bar{\epsilon}} = \begin{bmatrix} \epsilon_{xx} & \epsilon_{xy} & 0 \\ \epsilon_{yx} & \epsilon_{yy} & 0 \\ 0 & 0 & \epsilon_{zz} \end{bmatrix}; \bar{\bar{\mu}} = \begin{bmatrix} \mu_{xx} & \mu_{xy} & 0 \\ \mu_{yx} & \mu_{yy} & 0 \\ 0 & 0 & \mu_{zz} \end{bmatrix}.
$$

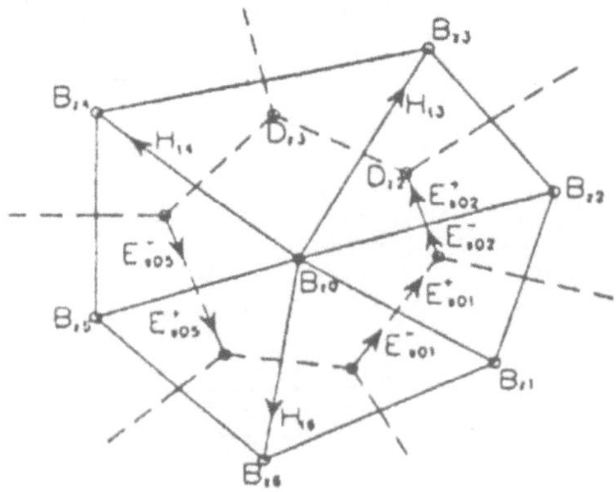

FIG. 5.1. *Anisotropic Media*

The case of H-polarization (TE) will be detailed. The case of E-polarization (TM) may be treated similarly by invoking duality [14]. Thus,

$$\vec{H} = H_z \hat{z},\ \vec{B} = \bar{\bar{\mu}}\vec{H} = B_z \hat{z},\ \vec{D} = \bar{\bar{\epsilon}}\vec{E},\ \vec{J} = \bar{\bar{\sigma}}\vec{E}. \tag{5.4}$$

Equations (5.1) and (5.2) may be combined to yield

$$\oint_{\partial D} \bar{\bar{\epsilon}}^{-1} \cdot \nabla \times (H_z \hat{z}) \cdot d\vec{l} = \int\int_D \omega^2 \mu_{zz} H_z \, dA \tag{5.5}$$

where $\bar{\bar{\epsilon}}$ now includes the elements of the conductivity tensor. Defining

$$C_{ij} = \epsilon_{ij}/d,\ d = \epsilon_{xx}\epsilon_{yy} - \epsilon_{xy}\epsilon_{yx} \tag{5.6}$$

this becomes

$$\oint_{\partial D} \hat{n} \cdot \bar{\bar{C}}^T \cdot \nabla H_z \, dl = -\int\int_D \omega^2 \mu_{zz} H_z \, dA, \tag{5.7}$$

while letting

$$\alpha = \hat{n} \cdot \bar{\bar{C}}^T \cdot \hat{n},\ \beta = \hat{n} \cdot \bar{\bar{C}}^T \cdot \hat{l} \tag{5.8}$$

produces

$$\oint_{\partial D} (\alpha \frac{\partial H_z}{\partial n} + \beta \frac{\partial H_z}{\partial l}) \, dl = -\int\int_D \omega^2 \mu_{zz} H_z \, dA. \tag{5.9}$$

Upon discretization, we arrive at

$$\sum_{i=1}^{m} [(\alpha_i^+ \frac{\partial H_z}{\partial n_i} + \beta_i^+ \frac{\partial H_z}{\partial l_i})\Delta l_i^+ + (\alpha_i^- \frac{\partial H_z}{\partial n_i} + \beta_i^- \frac{\partial H_z}{\partial l_i})\Delta l_i^-] \tag{5.10}$$

$$= -\omega^2 \sum_{i=1}^{m} (\mu_{zz}^+ A_i^+ + \mu_{zz}^- A_i^-) H_{z_0}. \tag{5.11}$$

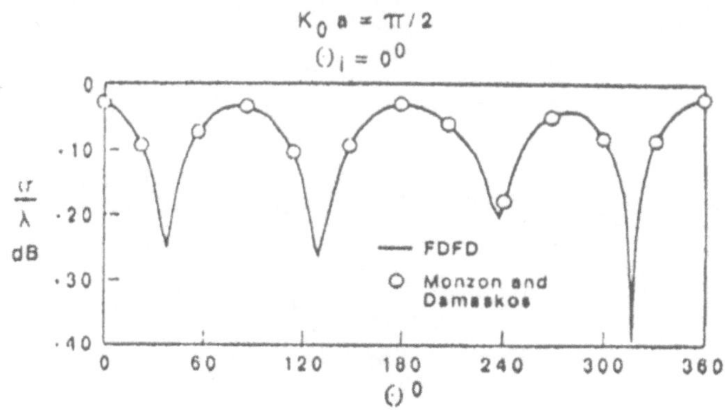

FIG. 5.2. *TE Anisotropic Rod*

In the above, the normal derivatives are directly approximated by central differences whereas the tangential derivatives require bilinear interpolation over the Delaunay triangulation prior to being so approximated.

In Figure 5.2, we compare our computed results with those of Monzon and Damaskos [15] for an anisotropic rod with tensor permittivity ($\mu_{zz} = 2\mu_0; \epsilon_{xy} = -\epsilon_{yx} = 2\epsilon_0; \epsilon_{xx} = \epsilon_{yy} = 4\epsilon_0$) embedded in free- space and illuminated by an H-polarized plane wave incident from $\theta = 0$. The scattered field in the form of the bistatic cross-section is presented with the circles representing the published series solution and the solid curve resulting from our numerical simulation.

6. Three-dimensional frequency-domain simulation. In three dimensions, the Dirichlet tessellation is composed of polytopes and the Delaunay tessellation consists of tetrahedra (Figure 6.1). Moreover, duality is preserved in the sense that edges of one tessellation are orthogonal to the corresponding faces of the other tessellation (Figure 6.2). Thus, the Control Region Approximation is directly extendable to three dimensions. In the remainder of this section, we present such an extension to Maxwell's equations in the frequency- domain [16]. The corresponding extension in the time-domain is then immediate.

We begin with the time-harmonic Faraday's law and Ampere's law, respectively,

$$\nabla \times \vec{E} = +j\omega\mu \vec{H} \tag{6.1}$$

$$\nabla \times \vec{H} = -j\omega\epsilon \vec{E} \tag{6.2}$$

together with the Leontovitch boundary condition

$$\vec{E} - (\vec{E} \cdot \hat{n})\hat{n} = \eta_s \, \hat{n} \times \vec{H} \tag{6.3}$$

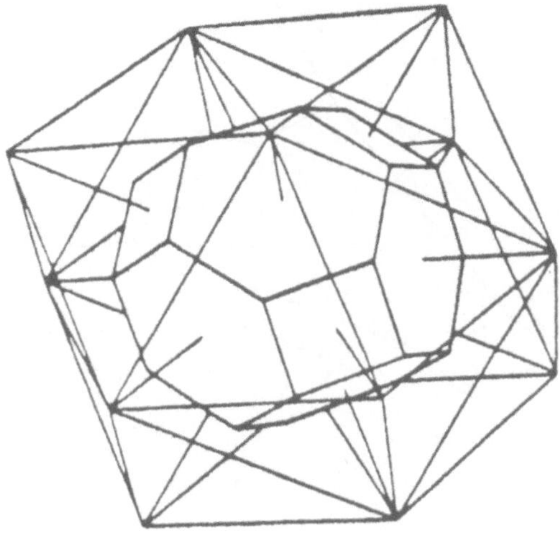

FIG. 6.1. *Three-Dimensional Tessellations*

along the surface of any conductor and the Sommerfeld radiation condition that all the components of the scattered fields, u, must, in the far field, behave according to

$$u_r - jku = o(1/r). \tag{6.4}$$

The grid points are distributed so that any Delaunay tetrahedron lies completely within a single material region and the material properties are assumed constant within each tetrahedron. The unknown field quantities of the computation are the component of $\vec{E}$ tangential to each tetrahedral edge and the component of $\vec{H}$ normal to each tetrahedral face (Figure 6.3). Invoking duality, this is equivalent to solving for the component of $\vec{E}$ normal to each polyhedral face and the component of $\vec{H}$ tangential to each polyhedral edge.

We now apply Stokes' theorem in order to recast Maxwell's equations in their natural circuital form

$$\oint E_t dl = + \int\int j\omega\mu H_n \, dS \tag{6.5}$$

$$\oint H_t dl = - \int\int j\omega\epsilon E_n \, dS \tag{6.6}$$

where Faraday's law is enforced on the faces of the Delaunay tetrahedra while Ampere's law is enforced on the faces of the Dirichlet polyhedra.

The radiation condition is approximated by a local absorbing boundary condition based upon truncation of the expansion of the scattered field into

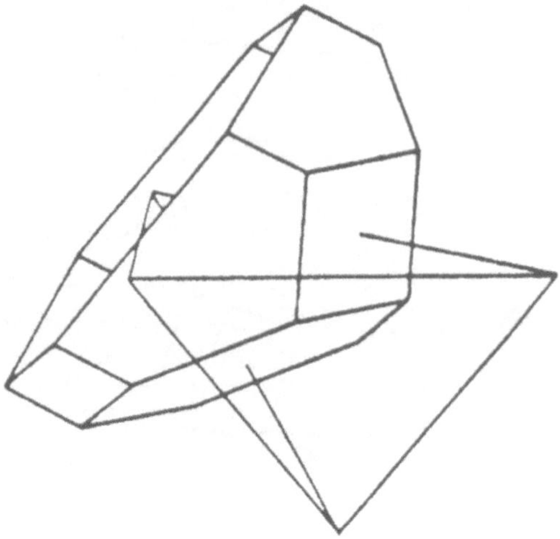

FIG. 6.2. *Duality*

outgoing waves. Specifically,

$$u \sim \frac{e^{jkr}}{r}\left[a_0(\theta,\phi) + \frac{a_1(\theta,\phi)}{r}\right], \tag{6.7}$$

where u is any field component. This expression is used to extrapolate the scattered fields at the outer computational boundary thus providing the required field values for the Control Region Approximation.

The above discretization procedure results in the system of equations

$$Au = b \tag{6.8}$$

where A is large and extremely sparse. The equations are first reordered so as to minimize fill-in during the subsequent elimination procedure. The matrix is next factored as $A = LU$ which is then followed by forward substitution $Lv = b$ and backward substitution $Uu = v$. This procedure is particularly efficient in the case of multiple incidence angles since only b is dependent on the angle of incidence. As a result, the expensive factorization phase need only be done once followed by repeated forward and backward substitutions which are relatively inexpensive.

If the EM field has been calculated for a fixed frequency, ω, then broad-band information can be obtained via the following procedure. The fields at neighboring frequencies are given by the Taylor series

$$u(\omega + \Delta\omega) = \sum_{n=0}^{\infty} \frac{(\Delta\omega)^n}{n!} u^{(n)}(\omega). \tag{6.9}$$

The successive derivatives of the field may be calculated from

$$Au^{(n)} = w^{(n)} \tag{6.10}$$

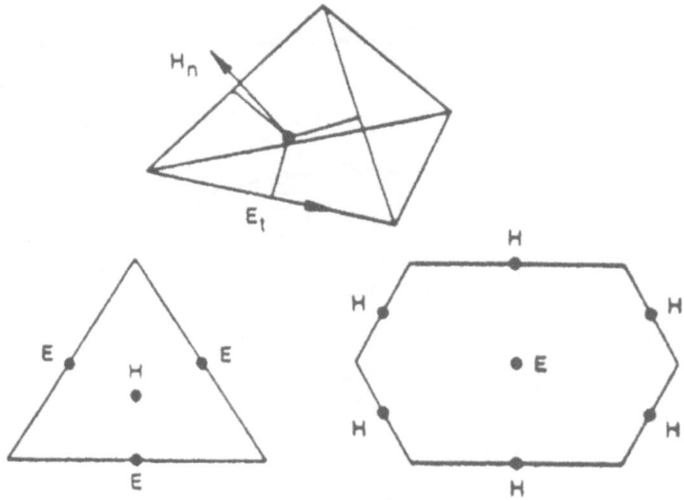

FIG. 6.3. *Control Region Approximation*

where

$$w^{(n)} := b^{(n)} - \sum_{k=1}^{n} \binom{n}{k} A^{(k)} u^{(n-k)} \tag{6.11}$$

with $A^{(k)}$ diagonal. Since A has already been factored, this broadband information is economically obtained. The above procedure may be improved by using Padé approximations in place of the Taylor series.

The quantity of ultimate interest in many applications is the radar cross-section (RCS) which for an incident field of

$$\vec{E}_i = \vec{A}\, e^{jk\vec{r}\cdot\hat{u}} \tag{6.12}$$

is given by

$$\sigma = 4\pi \left|\vec{F}_t(\hat{u}')\right|^2 / \left|\vec{A}\right|^2 \tag{6.13}$$

where

$$\begin{aligned} \vec{F}_t(\hat{u}') = \; & \tfrac{jk}{4\pi}\Big\{\hat{u}' \times \int\!\!\int (\hat{u}_n \times \vec{E}_s) e^{-jk\vec{r}\cdot\hat{u}'} dS \\ & -\eta_0 \hat{u}' \times \left[\hat{u}' \times \int\!\!\int (\hat{u}_n \times \vec{H}_s) e^{-jk\vec{r}\cdot\hat{u}'} dS\right]\Big\} \end{aligned} \tag{6.14}$$

is the far-field radiation pattern. In the above, $\hat{u}$ and $\hat{u}'$ are unit vectors in the directions of incidence and observation, respectively.

Figure 6.4 displays the normalized RCS for a PEC cube under plane wave illumination at broadside incidence. The grid point locations in this

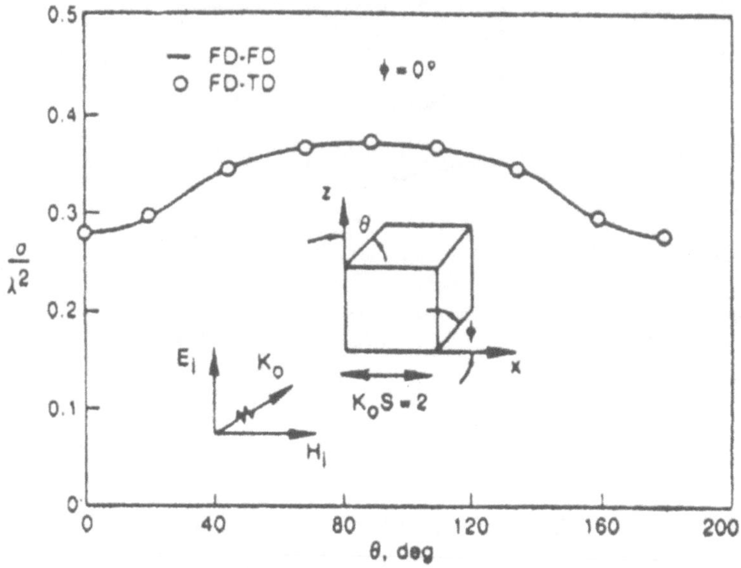

FIG. 6.4. *RCS of PEC Cube*

example form a body centered cubic lattice so that the resulting control regions are truncated octahedra not cubes. The electrical size of the cube is $k_0 s = 2$ where s is the cube edge length. A 20×20 mesh density was used on each face of the cube and the outer computational boundary was placed 15 cells from the surface of the cube. These numerical results are seen to be in close agreement with the FD-TD results of [17] where it is further shown that these results are in close agreement with those of a moment method calculation.

7. Two-dimensional periodic structures. We now provide a frequency domain analysis of the scattering of electromagnetic or acoustic waves by a two-dimensional periodic structure (Figure 7.1). This capability then allows one to analyze and design structures that either absorb the incident energy or diffract the incident field in preferential directions. There are ubiquitous applications of such structures in acoustics, optics, and microwave electronics. These applications include their use as filters, couplers, deflectors, modulators, and transducers in integrated optics and surface acoustics. The formulation to be presented permits any type of isotropic material including dielectrics, magnetic materials, and metallic conductors both with and without loss.

With reference to Figure 7.1, we assume a time- harmonic plane wave excitation and adhere to the convention $e^{-j\omega t}$ for all temporal variations. The incidence angle is measured from the normal to the structure with the clockwise direction being positive. The structure under consideration is

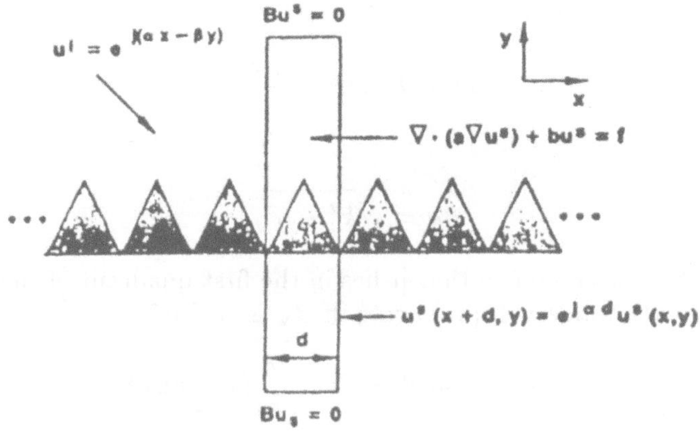

FIG. 7.1. *Periodic Structure*

infinite in the z-direction with no z-variation and infinite in the x-direction with period d. The equi-phase planes of the incident field are assumed parallel to the z-axis, i.e. the wave-vector, k, is orthogonal to the z-axis. Free-space conditions are assumed to prevail above the structure while below the structure is some homogeneous substrate (metallic, dielectric, magnetic, or free-space). Elsewhere within the unit cell, the structure is completely general with regards to its geometrical and material constituents except that any dielectric or magnetic materials are assumed to be isotropic.

Just as in Section 3, this problem reduces to a generalized Helmholtz equation in a cross-section orthogonal to the axis of the grating [20]:

$$\nabla \cdot (a \nabla u) + bu = 0 \tag{7.1}$$

the difference now being in the outer boundary conditions.

We assume that the incident field is a plane wave:

$$u_I = e^{j(\alpha x - \beta y)} \tag{7.2}$$

where

$$\alpha = k_0 \sin\theta, \beta = k_0 \cos\theta, k_0 = \omega\sqrt{\mu_0 \epsilon_0}. \tag{7.3}$$

This assumption on the incident field together with the periodicity of the structure yields the following pseudo-periodic boundary condition for the scattered field [18]:

$$u_S(x + d, y) = e^{j\alpha d} u_S(x, y). \tag{7.4}$$

Above the structure, the field has the following Rayleigh-Bloch expansion [19]:

$$u(x, y) = e^{j(\alpha x - \beta y)} + \sum_{N=-\infty}^{\infty} R_N e^{j(\alpha_N x + \beta_N y)} \tag{7.5}$$

where

$$\alpha_N = k_0 \sin\theta + \frac{2\pi N}{d} \tag{7.6}$$

and

$$\beta_N = \sqrt{k_0^2 - \alpha_N^2}. \tag{7.7}$$

The sign of β_N is chosen so that it lies in the first quadrant of the complex plane. The Nth mode will propagate iff β_N is real iff

$$-\frac{d}{\lambda_0}(1 + \sin\theta) < N < \frac{d}{\lambda_0}(1 - \sin\theta). \tag{7.8}$$

Below the structure, the field has the Rayleigh-Bloch expansion:

$$u(x,y) = \sum_{N=-\infty}^{\infty} T_N e^{j(\alpha_N x - \beta_N y)} \tag{7.9}$$

where α_N is as before and

$$\beta_N = \sqrt{k_1^2 - \alpha_N^2}. \tag{7.10}$$

The sign of β_N is chosen as before and again the Nth mode will propagate iff β_N is real iff

$$-\frac{d}{\lambda_0}(\frac{k_1}{k_0} + \sin\theta) < N < \frac{d}{\lambda_0}(\frac{k_1}{k_0} - \sin\theta) \tag{7.11}$$

where k_1 is the wave number in the substrate.

In either case, if we go far enough away from the structure either above or below we can ignore the exponentially decaying evanescent fields and use the above expansions to derive an outgoing radiation boundary condition [21]. We thus truncate the computational domain a finite distance above and below the structure where we represent the field as

$$u(x,y) = a_1(x)e^{j\beta_1 y} + \ldots + a_n(x)e^{j\beta_n y}. \tag{7.12}$$

For example, we represent the field at a fixed x-location along the upper boundary in terms of propagating modes:

$$u(y) = \sum_{i=1}^{n} a_i e^{j\beta_i y} \tag{7.13}$$

and interpolate at n equally spaced points along the vertical direction:

$$u_k = \sum_{i=1}^{n} c_i^k \alpha_i \; (k = 0, \ldots, n-1), \tag{7.14}$$

where, with $\Delta =$ vertical spacing,

$$\alpha_i := a_i e^{j\beta_i y_0}, \quad c_i := e^{-j\beta_i \Delta}. \tag{7.15}$$

This system of equations is written as:

$$C\vec{\alpha} = \vec{u}. \tag{7.16}$$

We then differentiate our modal expansion along the boundary:

$$u'(y_0) = j\vec{\beta} \cdot \vec{\alpha} = j\vec{\beta} \cdot (C^{-1}\vec{u}), \tag{7.17}$$

which can be rewritten as:

$$u'(y_0) = j\vec{\xi} \cdot \vec{u}; \; \vec{\xi} := C^{-T}\vec{\beta}. \tag{7.18}$$

Thus, we first solve the interpolation problem $C^T\vec{\xi} = \vec{\beta}$ and then employ its solution, $\vec{\xi}$, in the absorbing boundary condition (ABC) $u'(y_0) = j\vec{\xi} \cdot \vec{u}$. Note that the coefficient matrix C^T is Vandermonde and hence the ABC coefficients can be solved for by an efficient procedure.

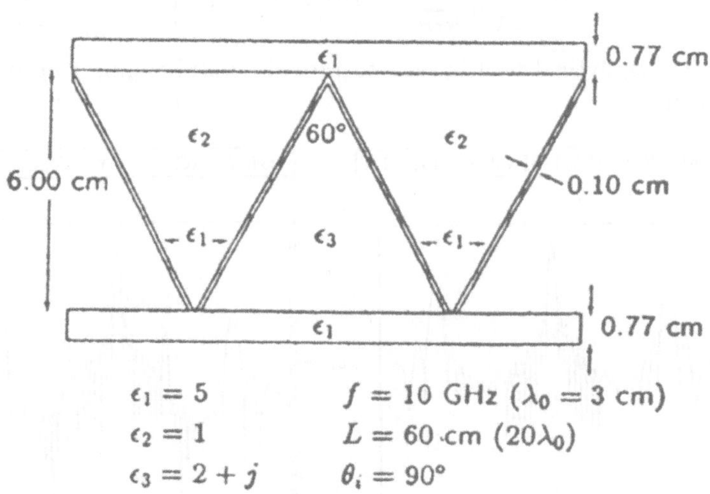

FIG. 7.2. *Webbed Structure*

Figure 7.2 displays a model of a radar-absorbing "webbed" structure suitable for jet engine cooling fin applications. The region defined by ϵ_1 provides structural support. The region defined by ϵ_2 is composed of the air cooling passages. The region defined by ϵ_3 is where the absorbent loading is placed to reduce the radar cross section (RCS). Both above and below the structure are free-space regions.

At a frequency of 10GHz and normal incidence, this structure supports five propagating modes. We can calculate the TM far-field behavior for an incident beam of finite width, L, [22] (Figure 7.3)

$$S(\theta) = kL^2 \left| \sum_{N \epsilon P} \frac{\sin [kL(\sin\theta - \sin\theta_N)/2]}{kL(\sin\theta - \sin\theta_N)/2} R_N(\theta_i) \right|^2 \sin\theta. \tag{7.19}$$

For the TE case, the $\sin\theta$ term is absent.

The following two cases were analyzed:

- TM Lossless
  - perfect energy balance (B = .997)
  - prominent 60° sidelobes
- TM Lossy
  - energy absorbed (B = .474)
  - slightly reduces specular reflection (backscatter)
  - heavily suppresses 60° sidelobes
  - enhances 26° sidelobes

where the energy balance parameter is

$$B = \left\{ \frac{1}{\beta} \sum_{N \epsilon P} \beta_N \left[|R_N|^2 + |T_N|^2 \right] \right\}^{1/2}. \tag{7.20}$$

Thus, this analytical tool is seen to provide useful engineering design information.

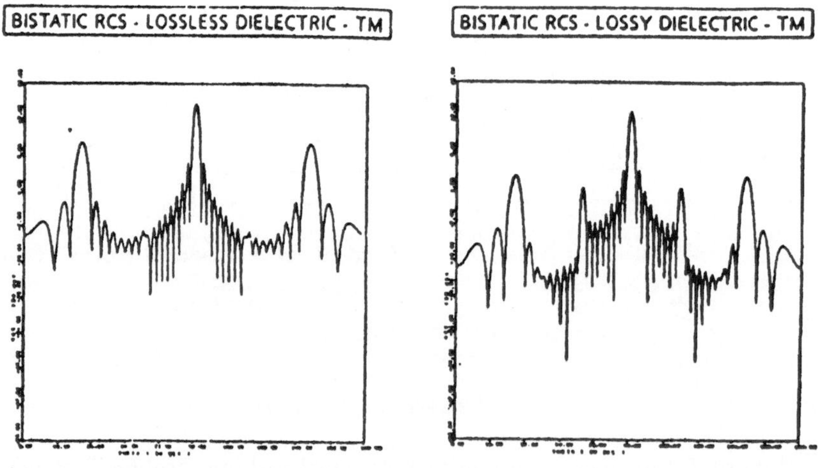

FIG. 7.3. *Bistatic RCS*

8. Conclusion. In the foregoing sections, the Control Region Approximation has been presented and shown to yield a physically meaningful discretization of the equations of electromagnetic scattering. Moreover, this technique accomodates completely general geometries and materials. Also, correct interfacial jump conditions are guaranteed. Boundary conditions both along the body and "at infinity" are enforced in a natural manner.

A number of applications of this procedure have been presented. A two-dimensional frequency-domain simulation of an airfoil has been compared to experimental data. A two-dimensional time-domain simulation of a cylinder coated with lossy magnetic material has been compared to the analytical solution. A two-dimensional anisotropic rod has been treated and compared to a published solution. A three-dimensional frequency-domain simulation of a cube has been compared to published time-domain results. Finally, a "webbed" structure modelling a jet engine cooling passage has been analyzed.

Lastly, ongoing work will now be outlined. Modelling of axisymmetric bodies reduces to the solution of two coupled two-dimensional Helmholtz-type equations and is hence amenable to this approach. Dielectric waveguides with periodic variation give rise to guided, leaky, and radiating modes which may be simulated as above. This analysis may then be used with the above work on periodic structures to predict the efficiency of periodic couplers. Also, the above analysis of two-dimensional periodic structures may be extended to three dimensions. Finally, the above analysis can be adapted to compute the propagating modes of cylindrical waveguides of arbitrary cross-section.

REFERENCES

[1] McCartin, B. J. and LaBarre, R. E., "Control Region Approximation: Preliminary Report", *Abstracts of the AMS*, Vol. 6, No. 3, June 1985, p. 266.

[2] McCartin, B. J. and LaBarre, R. E., "Generalized Control Region Approximation: Preliminary Report", *Abstracts of the AMS*, Vol. 6, No. 4, August 1985, p. 316.

[3] McCartin, B. J. and LaBarre, R. E., "Numerical Integration on Arbitrary Domains: Preliminary Report", *AMS*, Vol. 6, No. 6, November 1985, p. 486.

[4] McCartin, B. J. and LaBarre, R. E., "Control Region Approximation", *Proceedings of SIAM Spring Meeting*, Pittsburgh, PA, June 24–26 1985, p. A26.

[5] McCartin, B. J. and LaBarre, R. E., "Numerical Computation on Arbitrary Lattices of Points", *Proceedings of the International Congress of Mathematicians*, Berkeley, CA, August 3–11, 1986, p. LXXI.

[6] Caspar, J. R., Hobbs, D. E., and Davis, R. L., "Calculation of Two-Dimensional Potential Cascade Flow Using Finite Area Methods", *AIAA J.*, Vol. 19, No. 1, 1980, pp. 103–109.

[7] McCartin, B. J., "Discretization of the Semiconductor Device Equations", New Problems and New Solutions for Device and Process Modelling, Boole Press, 1985, pp. 72–80.

[8] McCartin, B. J., "Computational Electromagnetics", *Proceedings of the Centennial Celebration of the American Mathematical Society*, Providence, RI, August 8–12, 1988, p. 43.

[9] Kong, J. A., Electromagnetic Wave Theory, Wiley, 1986.

[10] Jones, D. S., Acoustic and Electromagnetic Waves, Oxford, 1986.

[11] McCartin, B. J., Bahrmasel, L. J., and Meltz, G., "Application of the Control Region Approximation to Two-Dimensional Electromagnetic Scattering", Chapter 5 of Differential Methods in Electromagnetic Scattering, M. A. Morgan (ed.), Elsevier, 1988.

[12] Lee, C. F., Shin, R. T., Kong, J. A.,and McCartin, B. J., "Absorbing Boundary Conditions on Circular and Elliptical Boundaries", *J. Electromagnetic Waves and Applications*, Vol. 4, No. 10, 1990, pp. 945–962.

[13] Lee, C. F., McCartin, B. J., Shin, R. T., and Kong, J. A., "A Triangular-Grid Finite-Difference Time-Domain Method for Electromagnetic Scattering Problems", *J. Electromagnetic Waves and Applications.*, Vol. 8, No. 4, 1994, pp. 449–470.

[14] Rappaport, C. M. and McCartin, B. J., "FDFD Analysis of Electromagnetic Scattering in Anisotropic Media Using Unconstrained Triangular Meshes", *IEEE Trans. Antennas and Propagation*, Vol. 39, No. 3, March 1991, pp. 345–349.

[15] Monzon, C. and Damaskos, N., "Two-Dimensional Scattering by a Homogeneous Anisotropic Rod", *IEEE Trans. Antennas and Propagation*, Vol. 34, 1986, pp. 1243–1249.

[16] McCartin, B. J. and DiCello, J. F., "Three Dimensional Finite Difference Frequency Domain Scattering Computation Using the Control Region Approximation", *IEEE Trans. Magnetics*, Vol. 25, No. 4, July 1989, pp. 3092–3094.

[17] Taflove, A. and Umashankar, K., "Radar Cross Section of General Three-Dimensional Scatterers", *IEEE Trans.*, Vol. EMC-25, 1983, pp. 433–440.

[18] Petit, R. (ed.), Electromagnetic Theory of Gratings, Springer-Verlag, 1980.

[19] Wilcox, C. H., Scattering Theory for Diffraction Gratings, Springer-Verlag, 1984.

[20] McCartin, B. J. and Kriegsmann, G. A., "Scattering by Two-Dimensional Periodic Structures", *J. Electromagnetic Waves and Applications*, to appear.

[21] Kriegsmann, G. A., "Radiation Conditions for Wave Guide Problems", *SIAM J. Sci. Stat. Comp.*, Vol. 3, No.3, Sep. 1982, pp. 318–326.

[22] Kriegsmann, G. A., "Scattering by Acoustically Large Corrugated Planar Surfaces", *J. Acoust. Soc. Am.*, Vol. 88, No.1, July 1990, pp. 492–495.

STRUCTURAL ACOUSTIC INTERACTIONS AND ON SURFACE CONDITIONS

MICHAEL J. MIKSIS* AND LU TING†

Abstract. The interaction of an acoustic wave and a clamped elastic panel is considered. Here we present numerical results comparing the exact solution found by solving the coupled acoustic wave - elastic panel problem to two decoupling approximations using an on surface boundary condition. These on surface conditions are derived in the limit where the ratio of the acoustic sound speed in the fluid to the surface wave speed in the panel is small.

1. Introduction. In order to study the interaction of an acoustic wave and an elastic panel, one needs to determine both the scattered field above and below the panel plus the dynamics of the panel. Hence the partial differential equations for the three-dimensional acoustic field must be solved coupled to the equations of motion of the panel, which is two-dimensional of a finite extent. An approximation, decoupling the acoustic field from the dynamics of the scatterer, has been recently suggested by Kriegsmann and Scandrett [1]. They applied the far field radiation conditions on the surface of a baffled membrane. By considering the limit where the acoustic wave length to the surface wave length is small, Miksis and Ting [2] were able to derive systematically the on surface conditions in [1] and gave an asymptotic derivation of the result in the time harmonic two-dimensional near normal incidence case. Another approach to deriving the on surface conditions on a baffled membrane was given by Kriegsmann and Scandrett [3]. They cast the problem as a dual integral equation and then approximated the equations kernel's. Later Kriegsmann and Scandrett [4] applied the on surface condition to the acoustic problem of scattering from an infinite array of baffled, fluid-loaded membranes. Miksis and Ting [5] extended their analysis for membranes to the problem of the interaction of an acoustic wave and a clamped elastic panel. They found by assuming that the ratio of the acoustic sound speed in the fluid to the sound speed in the panel is small, that they could derive systematically an on surface boundary condition. In particular, they found that the condition of near normal incidence, which was required in the membrane case [2] to justify the on surface condition, could be relaxed. Ting [6] has recently extended the analysis in Miksis and Ting [5] to the case of moving media. Our aim here, is to present numerical examples of the interaction of an acoustic wave and a clamped elastic plate. We will compare the exact numerical answer which keeps all the coupling in the problem with the asymptotic analysis

* Department of Engineering Sciences and Applied Mathematics, Northwestern University, Evanston, Illinois 60208.

† Courant Institute of Mathematical Sciences, New York University, New York, NY 10012.

using the on surface conditions derived by Miksis and Ting [5].

2. Formulation. Consider the scattering of an acoustic wave by a planar interface, the z x plane. The interface is a rigid surface except on the flexible panel, $\bar{\mathcal{D}}$. Let its boundary be denoted by $\mathcal{C}$ and its interior by $\mathcal{D}$. We consider the case where the ratio of the characteristic acoustic length ℓ, e.g., the wave length or inverse wave number, to the characteristic surface scale L is small, i. e.,

$$\epsilon = \ell/L \ll 1 . \tag{2.1}$$

Using ϵ as the small expansion parameter, we shall derive systematically an *on surface condition* which relates the acoustic pressure on one side of the interface to the surface deformation. This derivation can be found in Miksis and Ting [5] but is presented here for completeness. Using the results of the derivation, a closed system of equations for the panel oscillation including the second order effect of the acoustic field can be determined. Hence we can solve for the panel oscillation first and then write down the solutions for the acoustic field in terms of the panel oscillation.

Let the velocity be scaled by the speed of sound, C, the density by the ambient density, ρ_0, and the length by ℓ. The size of the panel $\bar{\mathcal{D}}$ is $O(L) = O(\ell/\epsilon)$. Consider the acoustic field induced by an incident wave, $\phi^{(i)}$. The velocity potential $\Phi(t, x, y, z)$ is governed by the simple wave equation,

$$(\partial^2_{tt} - \partial^2_{xx} - \partial^2_{yy} - \partial^2_{zz})\, \Phi = 0 \ , \quad \text{for} \quad y > 0 \ . \tag{2.2}$$

Note that $\phi^{(i)}$ is a solution of (2) in the whole space. Let $\bar{\phi}$ represent the reflected wave in the upper half space with a rigid panel, i. e., $\bar{\phi}(t, x, y, z) = \phi^{(i)}(t, x, -y, z)$ for $y \geq 0$. Due to the flexibility of the panel $\bar{\mathcal{D}}$, there is an addition contribution ϕ to the reflected wave. Thus we write $\Phi = \phi^{(i)} + \bar{\phi} + \phi$, and note that ϕ is also governed by the simple wave equation (2.2). The kinematic condition on the interface $y = 0$ is

$$\partial_y \phi = \partial_t \eta(t, x, z) \ , \tag{2.3}$$

where $\eta(t, x, z)$ denotes the vertical displacement of the interface from the xz plane, i. e., $\eta = 0$ for $(x, z) \notin \mathcal{D}$. For a clamped panel, we have $\partial_n \eta = 0$ and hence $\partial_x \eta = 0$, and $\partial_z \eta = 0$ on $\mathcal{C}$, in addition to $\eta = 0$ on $\mathcal{C}$, where ∂_n denotes the inward normal derivative on $\mathcal{C}$. If the front of the incident wave hits the panel at $t = 0$, we can impose the homogeneous conditions on ϕ and η,

$$\phi = 0 \ , \ \partial_t \phi = 0 \quad \text{and} \quad \eta = 0 \ , \ \partial_t \eta = 0 \qquad \text{for} \quad t \leq 0 \ . \tag{2.4}$$

The solution of (2.2) subject to the above conditions is given by the Kirchhoff formula,

$$\phi(t, x, y > 0, z) = -\frac{1}{2\pi} \int\!\!\int_{\bar{\mathcal{G}}} \frac{\eta_t(t - R, x', z')}{R}\, dx'dz' \ , \tag{2.5}$$

where $R = [(x-x')^2 + y^2 + (z-z')^2]^{1/2}$ denotes the distance from the point $P(x, y>0, z)$ in the upper half plane to a source at $(x', 0, z')$ created at the retarded time $t-R$. The domain of dependence of $\phi(t,x,y,z)$ is the circular disc $\bar{\mathcal{H}}$ in the $x'z'$ plane, i.e., $\{\bar{\mathcal{H}} \mid R \le t\}$ or, $r^2 = (x'-x)^2 + (z'-z)^2 \le t^2 - y^2$. The domain of integration in (2.5), is the intersection of $\bar{\mathcal{H}}$ and the panel, i. e., $\bar{\mathcal{G}} = \bar{\mathcal{H}} \cap \bar{\mathcal{D}}$. The integrand becomes singular as y and $r \to 0^+$. To remove this singularity, we introduce the polar coordinates, r, θ, centered at (x, z), i. e., $x' - x = r\cos\theta$ and $z' - z = r\sin\theta$ and (2.5) becomes,

$$\phi(t, x, y>0, z) = -\frac{1}{2\pi}\int\left[\int_{\bar{\mathcal{G}}} d\theta\; g_t(t-R, r, \theta)\right]\frac{r\,dr}{R}, \tag{2.6}$$

with $g(t,r,\theta) = \eta(t, x + r\cos\theta, z + r\sin\theta)$ and $R = (r^2 + y^2)^{1/2}$. Using (2.3) and (2.4) and letting $y \to 0^+$ and $R \to r$ in (2.6) we find the iterated integral for the acoustic pressure induced by the panel oscillation,

$$-\phi_t(t, x, 0^+, z) = \frac{1}{2\pi}\int_0^{2\pi}\int_0^{t} g_{tt}(t-r, r, \theta)\; dr\, d\theta\,. \tag{2.7}$$

Due to symmetry, the negative transmitted pressure on $y \to 0^-$ is also given by the right hand side of (2.7). With the acoustic load on the panel related to the panel oscillation, the coupled system of equations for the acoustic/panel interaction problem is reduced to an integro-differential equation for the panel oscillation (see eq. (4.4) for the 2-D time harmonic version of this). In Section **3** we shall reduce the nonlocal integral relationship (2.7) to a local one for the limiting case (2.1).

3. On surface condition. To make use of (2.1), we change the spatial variables x, z of the surface deformation η to those scaled by L, i. e., $\eta = \tilde{\eta}(t, \tilde{x}, \tilde{z})$ with $\tilde{x} = \epsilon x$ and $\tilde{z} = \epsilon z$. In terms of the integration variables in (2.7), we have $g(t, r, \theta) = \tilde{\eta}(t, \tilde{x}', \tilde{z}')$ with $\tilde{x}' = \tilde{x} + \epsilon r\cos\theta$ and $\tilde{z}' = \tilde{z} + \epsilon r\sin\theta$. We now use integration by parts to expand the integral in (2.7) in powers of ϵ.

Using (2.4) and (2.7), we have at $r = t$ that $g_t(0, t, \theta) = 0$ and at $r = 0$ that $g_t(t, 0, \theta) = \tilde{\eta}_t(t, \tilde{x}, \tilde{z})$. Integrating by parts and using the conditions along $y = 0$ we arrive at

$$\phi_t(t, x, 0^+, z) = -\tilde{\eta}_t(t, \tilde{x}, \tilde{z}) - \frac{1}{2\pi}\int_0^{2\pi} d\theta\; g_r(t, 0, \theta) - \tag{3.1}$$

$$-\frac{1}{2\pi}\int_0^{2\pi}\int_0^{t} g_{rr}(t-r, r, \theta)\; dr\, d\theta\,.$$

Note that as a point crosses over the boundary $\mathcal{C}$ along its inward normal, η_{nn} changes from 0 to the limiting value from the interior of $\bar{\mathcal{D}}$, therefore, g_{rr} is a step function across $\mathcal{C}$.

To remove the second term on the right-hand side of (3.1) and similar terms later in the higher order approximations, we note that

$$\partial_{r^n} g(t,r,\theta) = \epsilon^n [\hat{r}\cdot\tilde{\nabla}]^n \tilde{\eta}(t,\tilde{x}',\tilde{z}'), \qquad n = 1,2,\dots, \tag{3.2}$$

where $\hat{r}\cdot\tilde{\nabla} = \cos\theta\ \partial_{\tilde{x}'} + \sin\theta\ \partial_{\tilde{z}'}$. By the method of induction we can express the θ-average of the left hand side of (3.2) at $r = 0$, I_n, in terms of local derivatives of $\tilde{\eta}$. For an odd $n = 2m+1$ and for an even $n = 2m$ respectively, we have

$$I_{2m+1} = 0 \quad \text{and} \quad I_{2m} = \frac{(2m)!\epsilon^{2m}}{2^{2m}\ m!\ m!}\tilde{\Delta}^m \tilde{\eta}(t,\tilde{x},\tilde{z}) , \tag{3.3}$$

where $\tilde{\Delta} = \partial^2_{\tilde{x}\tilde{x}} + \partial^2_{\tilde{z}\tilde{z}}$. For $n = 1$, (3.3) says that the second term on the right-hand side of (3.1) equals zero and (3.1) becomes

$$\phi_t(t,x,0^+,z) = -\tilde{\eta}_t(t,\tilde{x},\tilde{z}) - \frac{1}{2\pi}\int_0^{2\pi}\int_0^t g_{rr}(t-r,r,\theta)\ dr\ d\theta . \tag{3.4}$$

Since $r \le t$, the length of the interval(s) of integration in r is less than or equal to t so the last term in (3.4) is $O(t\epsilon^2)$, and (3.4) yields the classical plane wave approximation PWA, $\phi_t = -\tilde{\eta}_t + O(\epsilon^2 t)$. When we put back the length and time scales, the error is $O(\epsilon^2 Ct/\ell) = O(\frac{\ell}{L}\ \frac{Ct}{L})$

To get a higher order approximation we differentiate (3.4) with respect to t, integrate by parts, use (3.2) and (3.3), and obtain the next order on surface condition,

$$\begin{aligned}\phi_{tt}(t,x,0^+,z) &= -\tilde{\eta}_{tt}(t,\tilde{x},\tilde{z}) - \frac{\epsilon^2}{2}\tilde{\Delta}\tilde{\eta}(t,\tilde{x},\tilde{z}) \\ &\quad - \frac{\epsilon^2}{2\pi}\int_0^S \frac{H(t-r^*)ds}{r^*}[\hat{\tau}\cdot\hat{\theta}]^3\tilde{\Delta}\tilde{\eta}(t-r^*,\epsilon X,\epsilon Z) + O(t\epsilon^3) .\end{aligned} \tag{3.5}$$

where $\mathbf{r}^* = r^*\hat{r} = (X-x)\hat{\imath} + (Z-z)\hat{k}$ and $\hat{r} = \hat{k}\cos\theta + \hat{\imath}\sin\theta$. Here $X(s)$ and $Z(s)$ are the coordinates of the contour $\mathcal{C}$, s is arc length and the unit tangent vector is $\vec{\tau} = \vec{\imath}X'(s) + \vec{k}Z'(s)$.

Now we shall show that for $t = O(1)$ the contribution of the integral along the arc(s) of $\mathcal{C}$ to the panel oscillation is one order higher. The third term on the right-hand side of (3.5) will be present only for (x,z) in a narrow boundary strip $\bar{\mathcal{B}}$ where the distance $d(x,z)$ to $\mathcal{C}$ is less or equal to t, i. e., $\{\bar{\mathcal{B}} \mid (x,z) \in \bar{\mathcal{D}}$ and $d(x,z) \le t\}$, with $d = \min\{[(x-X)^2 + (z-Z)]^2\}$ $0 \le$

$s \leq S$. We define the subdomain $\{\bar{\mathcal{Q}} \mid (x,z) \in \bar{\mathcal{D}} \text{ and } d(x,z) \geq t\}$, in which the third term in (3.5) is absent, i.e.,

$$(3.6)\quad \phi_{tt}(t,x,0^+,z) = -\tilde{\eta}_{tt}(t,\tilde{x},\tilde{z}) - \frac{\epsilon^2}{2}\tilde{\Delta}\tilde{\eta}(t,\tilde{x},\tilde{z}) + O(t\epsilon^3)\,, \quad (x,z) \in \bar{\mathcal{Q}}\,.$$

Since the size of $\bar{\mathcal{B}}$ is of the order of $t/L = O(\epsilon)$ relative to the size of $\bar{\mathcal{D}}$ or $\bar{\mathcal{Q}}$, we can use (3.6) as the on surface condition on $\bar{\mathcal{D}}$ for $t = O(1)$ and expect that the error of the panel oscillation η and of the induced acoustic wave ϕ to remain $O(t\epsilon^3)$.

By applying an on surface condition for the load on the panel, the equation for the panel oscillation is uncoupled from the equations of the acoustic field. In order to determine the panel oscillation η to $O(t\epsilon^3)$ we need to solve the normal stress equation for the panel deformation (e.g., equation (4.1) in the time harmonic case) coupled with (3.6) for ϕ, only at $y = 0^+$ and (x,z) in $\bar{\mathcal{D}}$. The solution of this system requires much less work than the solution of the panel acoustic interaction problem in three-dimensional space.

4. Time harmonic on surface condition. Suppose we consider the time harmonic case. Miksis and Ting [5] determined by a systematic asymptotic analysis the on surface condition in the limit where the ratio of the speed of sound in the liquid to the surface wave speed is small, i.e., $\epsilon \ll 1$. This justified in the time harmonic case the analysis of the previous section. Here we will present numerical results which compare the leading order (PWA) and second order (SOA) on-surface condition with the exact numerical solution of the problem. Only the two-dimensional case will be considered here.

Suppose the panel has length $2L$ and lies along $y = 0$ from $-L \leq x \leq L$. So $\mid x \mid > L$ represents a rigid interface. Assume that a time harmonic wave of frequency ω strikes the x-axis. Let the unit normal to the incident wave front make an angle α with the x-axis. Hence $\alpha = \pi/2$ represents normal incidence. The dimensional normal stress equation of the panel deformation η is given by

$$(4.1)\qquad G\frac{d^4\eta}{dx^4} - \sigma\omega^2\eta = -2i\omega\rho\phi - 2p^i.$$

Here G is the flexural rigidity of the panel, σ is the panel area density, ρ is the fluid density and p^i is the pressure of the incident time harmonic wave. As noted in the previous section, equation (4.1) is coupled to the solution of the reduced wave equation in the region above the fluid. Hence as in the previous section, if we could relate ϕ to η we could reduce the problem down to solving only one equation.

Suppose we introduce a set of dimensionless coordinates where the unit of length is L, the unit of time is the inverse frequency and the unit of pressure is $\rho\omega^2L^2$. Then using (3.6) in (4.1) we find that to second order in ϵ, the surface deformation is given by [5]

$$\frac{d^4\eta}{dx^4} - \xi^2\left(\hat{k}+2i\gamma\right)\hat{k}^3\eta + i\gamma\xi^2\hat{k}\frac{d^2\eta}{dx^2} = -2\xi^2\hat{k}^4\gamma e^{i\hat{k}x\cos\alpha}. \tag{4.2}$$

Here we define $\xi = \epsilon C/C_m$, $C_m = \sqrt{G/\sigma L^2}$ as the surface wave speed and $\gamma = \rho L/\sigma$. The dimensionless wave number is defined as $\hat{k} = kL$ where k is the dimensional wave number. Note that if we set $\ell = k$, then $\epsilon = 1/kL = 1/\hat{k}$. So ϵ small requires $\hat{k}$ large. This scaling is different than that used by Miksis and Ting [5] in their systematic analysis of the time harmonic case but it is more convenient for numerical solutions. Also note that ξ was assumed to be order ϵ^2 in the analysis of Miksis and Ting [5], hence the product $\xi\hat{k}^2$ was assumed to be order one.

Solutions of equation (4.2) represent the second order approximation (SOA) for the surface deformation of the scattering of a time harmonic acoustic wave from an elastic panel. The leading order approximation, or Plane Wave Approximation (PWA), for the surface deformation can be found by using $\phi = -\eta$, see equation (3.4), as the on surface condition in (4.1). The result is

$$\frac{d^4\eta}{dx^4} - \xi^2\left(\hat{k}+2i\gamma\right)\hat{k}^3\eta = -2\xi^2\hat{k}^4\gamma e^{i\hat{k}x\cos\alpha}. \tag{4.3}$$

We see that the only difference between (4.2) and (4.3) is the $\frac{d^2\eta}{dx^2}$ term.

The problem can be formulated without any approximations. As we did in the time dependent case, the potential along the surface can be related to an integral of the surface deformation, see e.q., (2.5). This can then be substituted into (4.1) to produce an integral-differential equation for the surface deformation. The result is

$$\frac{d^4\eta}{dx^4} - \xi^2\hat{k}^4\eta = i\xi^2\hat{k}^4\gamma\int_{-1}^{1} H_0^1(\hat{k}R)\eta(\tilde{x})d\tilde{x} - 2\xi^2\hat{k}^4\gamma e^{i\hat{k}x\cos\alpha}. \tag{4.4}$$

Here $R = \mid x - \tilde{x} \mid$ and H_0^1 is a Hankel function.

We would like to compare the predictions from equations (4.2)-(4.4). We note that all of these equations should be solved with the boundary conditions of $\eta = \frac{d\eta}{dx} = 0$ at $x = \pm 1$. First we should identify the resonance frequencies. These can be determined by setting the right hand size of equation (4.4) to zero and solving. We find that the resonance frequencies occur at the roots β_i, $i = 1, 2, \ldots$ of the equation $\tanh(\beta) = \pm\tan(\beta)$ where $\beta^2 = \xi\hat{k}^2$. It is easy to show that two of the roots are $\beta_1 = 3.9266\ldots$ and $\beta_2 = 7.06858\ldots$. We expect that similar results will be predicted away from resonance for the exact (4.4), PWA (4.3) and second order approximation SOA (4.2). Hence we will consider only the resonance cases here.

Equations (4.2) and (4.3) are solved numerically by using an implicit Chebyshev pseudo-spectral method. The same discretization of the left hand side of (4.4) is used. The integral on the right hand side of (4.4) is also discretized at the Chebyshev collocation points but care must be taken because of the logarithmic singular term. The log singularity is accounted for by interpolating the smooth part of the integrand between collocation points and then integrating exactly. The regular integrals are discretized using the trapezoidal rule. The resulting matrix equation for η is then solved numerically.

Suppose we set $\xi = 0.25$, $\hat{k} = 2\beta_1$, $\gamma = 1$ and consider the normal incidence case. The predictions of the PWA, SOA and exact formulation for the magnitude of the surface displacement η are identical graphically. Hence in Figure 1 we plot the magnitude of the surface displacement as a function of x for the PWA, SOA and exact formulation. In Figure 1 the near normal case $\alpha = 2.9\pi/6$ is considered with the other parameters as above. Note that the SOA and exact solution are extremely close while the PWA is slightly off but still good. In Figure 2 we set $\alpha = \pi/3$ and again see that all the approximations are good but again the SOA is better than the PWA. In Figure 3 we again set $\xi = 0.25$, $\gamma = 1$ and $\alpha = \pi/3$ but now we consider $\hat{k} = 2\beta_2$. As with the previous plots, the SOA compares well with the exact solution. The effect of γ is illustrated in Figure 4 where we set $\gamma = 0.01$ and let the other parameters be the same as in Figure 3. Except for the difference in the predicted amplitude of η, the plots are similar to Figure 3. Suppose we increase the value of ξ. In Figure 5 we set $\xi = 0.5$, $\hat{k} = \sqrt{2}\beta_1$, $\gamma = 1$ and $\alpha = \pi/3$. Note that the SOA is still a better approximation than the PWA but differences are greater than in the smaller ξ case of Figure 2. This is to be expected since Miksis and Ting [5] have shown the vality of the approximation in the limit of small ξ (or ϵ). Hence as ξ increases, we can expect the approximations, both the PWA and SOA, to break down.

In conclusion, we see that the second order on surface condition gives a much better approximation than the plane wave approximation. Higher order approximations can be found by following the analysis presented in Section 3 or in Miksis and Ting [5]. Numerical results have only been presented in Section 4 for the two-dimensional elastic plate scattering problem. We expect similar conclusions for the three-dimensional case. The 3-D time harmonic case could be studied in a manner similar to how we did the 2-D case. The 3-D PWA and SOA model equations are partial differential equations but are easily solved. The exact 3-D solution can again be formulated as an integral-differential equation over the plate surface. Hence the amount of computational time to solve the exact problem now increases considerably because of the extra dimension. So the PWA and SOA approximations are useful in that they allow for computationally fast solutions. This is especially true in the time dependent case formulated in Section 3.

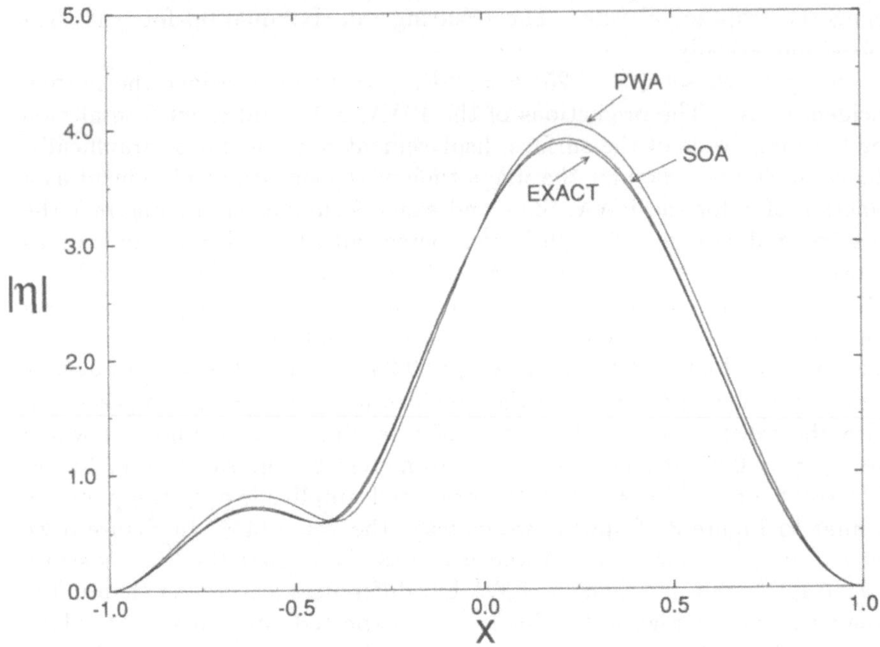

FIG. 1. *The magnitude of the surface deformation η as a function of x. We plot the solution of the PWA equation (4.3), the SOA equation (4.2) and the exact equation (4.4). Here we have set $\xi = 0.25$, $\hat{k} = 2\beta_1$, $\gamma = 1$ and $\alpha = 2.9\pi/6$.*

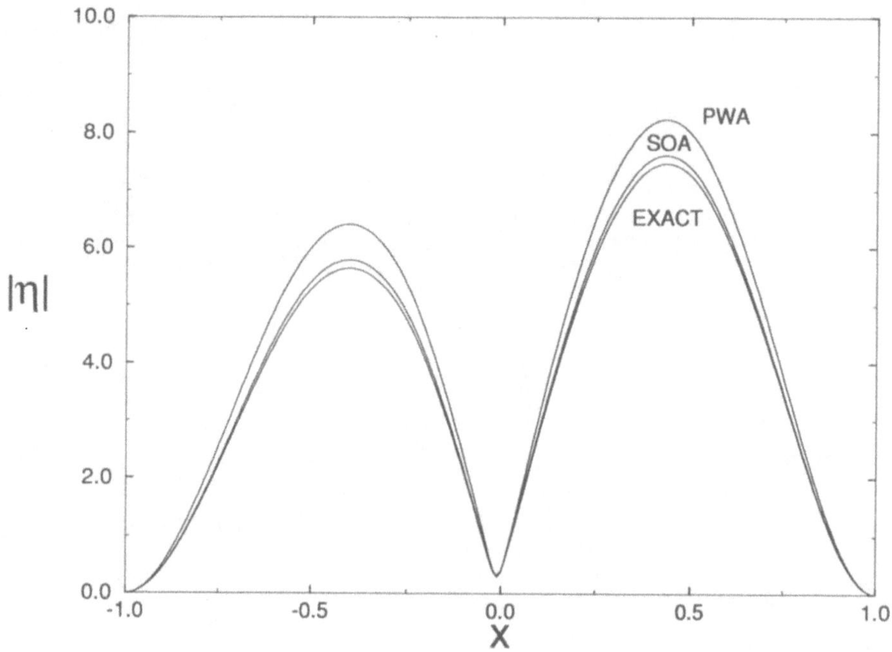

FIG. 2. *The magnitude of the surface deformation* η *as a function of* x *for* $\xi = 0.25$, $\hat{k} = 2\beta_1$, $\gamma = 1$ *and* $\alpha = \pi/3$.

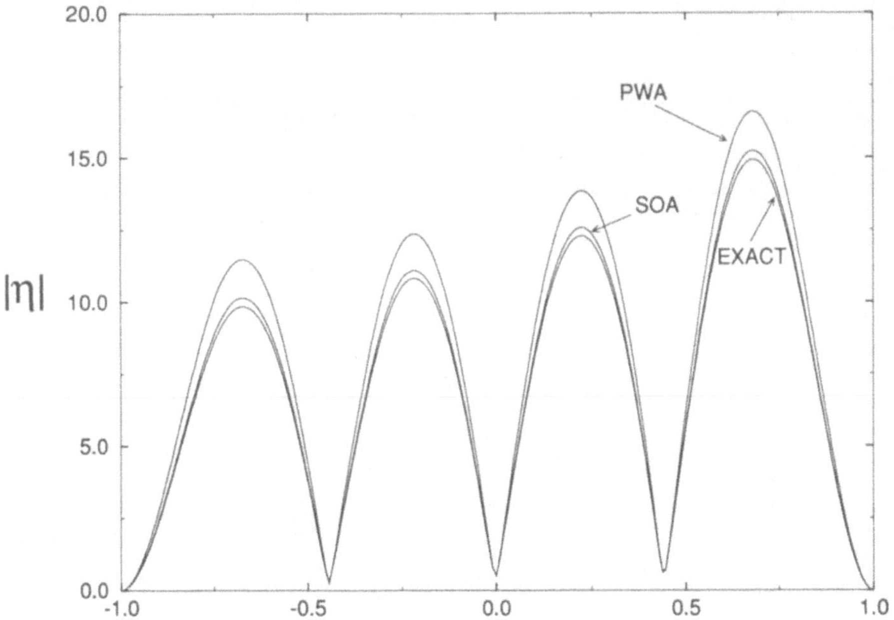

FIG. 3. *The magnitude of the surface deformation* η *as a function of* x *for* $\xi = 0.25$, $\hat{k} = 2\beta_2$, $\gamma = 1$ *and* $\alpha = \pi/3$.

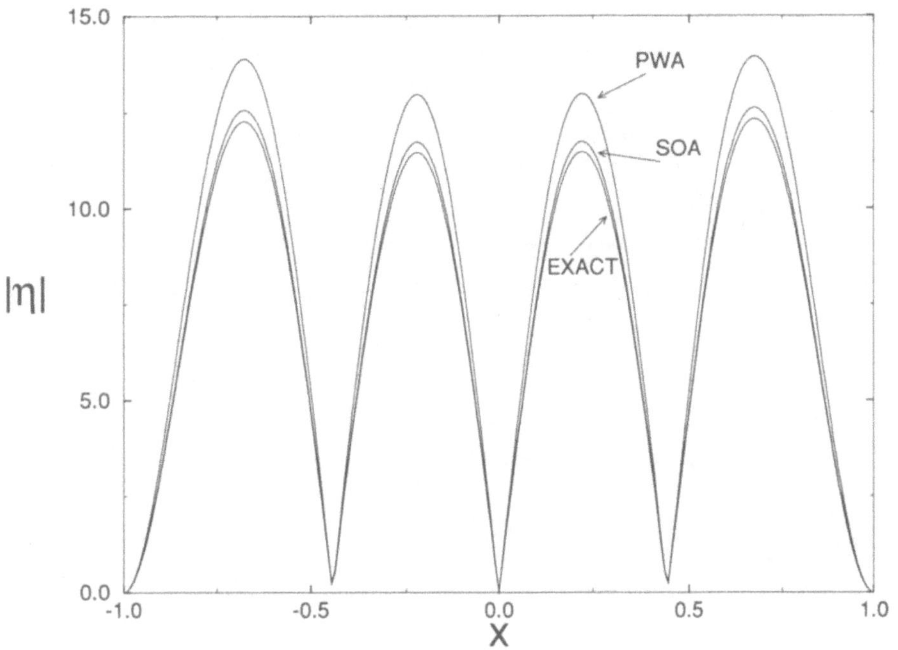

FIG. 4. *The magnitude of the surface deformation* η *as a function of* x *for* $\xi = 0.25$, $\hat{k} = 2\beta_2$, $\gamma = 0.01$ *and* $\alpha = \pi/3$.

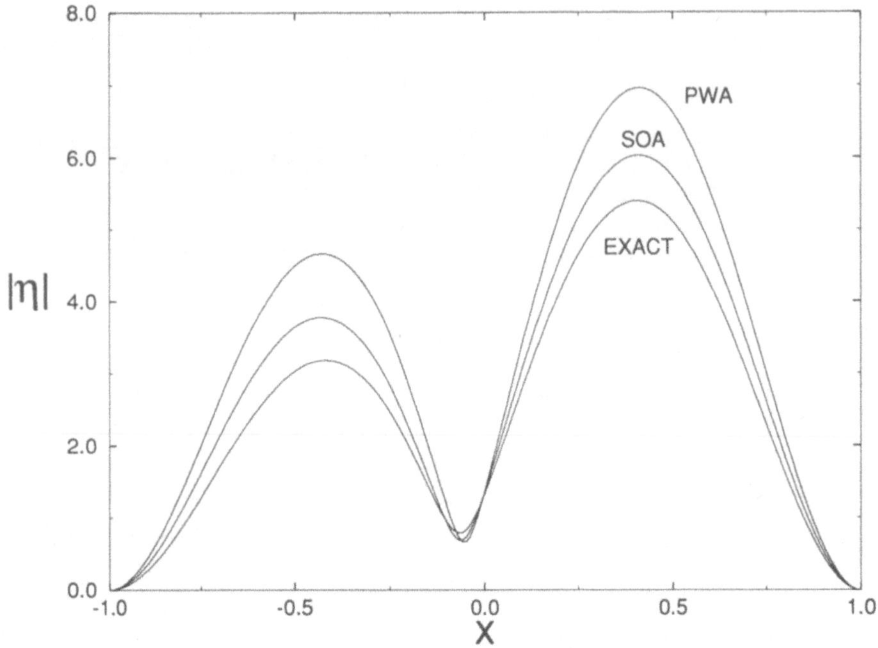

FIG. 5. *The magnitude of the surface deformation* η *as a function of* x *for* $\xi = 0.5$, $\hat{k} = \sqrt{2}\beta_1$, $\gamma = 1$ *and* $\alpha = \pi/3$.

Acknowledgments. MJM was supported in part by DOE Grant No. DE-FG02-88ER13927 and LT was supported in part by AFOSR Grant No. F49620-93-1-0027.

REFERENCES

[1] G.A. Kriegsmann and C.L. Scandrett, Assessment of a new radiation damping model for structural acoustic interactions, *J. Acoust. Soc. Amer.* **86** 788–794 (1989).

[2] M.J. Miksis and L. Ting, Scattering of an incident wave from an interface separating two fluids. *Wave Motion*, **11** 545–557 (1989).

[3] G.A. Kriegsmann and C.L. Scandrett, Decoupling Approximations for Structural Acoustic Interactions. *Appl. Math. Lett.* **3** 51–54 (1990).

[4] G.A. Kriegsmann and C.L. Scandrett, Decoupling Approximations Applied to an Infinite Array of Fluid Loaded Baffled Membranes. *J. Comp. Phys.* **111** 282–290 (1994).

[5] M.J. Miksis and L. Ting, Panel Oscillations and Acoustic Waves. *Appl. Math. Lett.*, **8** 37–42, (1995).

[6] L. Ting, On-Surface Conditions for Structural Acoustic Interactions in Moving Media. *SIAM J. Appl. Math.*, **55** 369–389 (1995).

WAVEFIELD REPRESENTATION USING COMPACT AND DIRECTIONALLY LOCALIZED SOURCES

ANDREW N. NORRIS* AND THORKILD B. HANSEN†

Abstract. New representations for compact source regions are developed. The wave field satisfies the homogeneous Helmholtz equation outside of some finite region, and is assumed known on some value of r =constant, possibly infinite. Given this data, we represent the radiated field by an equivalent distribution of sources in complex space. The point sources are located on the complex "sphere" $r = ia$, with a weighting uniquely defined by the far-field. Each complex point source acts like a Gaussian beam with a well defined directionality. This type of representation offers an alternative to the usual multipole expansion, and may be preferable if the source is highly directional. In the high frequency limit the far-field pattern function is proportional to the associated weighting function for the complex sources. The theory is developed for both 2D and 3D, and numerical examples are presented for a 3D end-fire array.

Key words. Complex points sources, Helmholtz equation, radiation

1. Introduction. We are concerned with solutions to the Helmholtz equation which are regular outside of a source region in R^d, $d = 2$ or 3. The sources are bounded by the surface S, and the exterior domain is V. The wave function $f(\mathbf{x})$ also satisfies the radiation condition. Thus,

$$\nabla^2 f(\mathbf{x}) + k^2 f(\mathbf{x}) = 0, \qquad \text{for } \mathbf{x} \text{ in } V, \tag{1.1}$$

and

$$\partial f/\partial r - ikf = O(f/r), \qquad r \to \infty, \tag{1.2}$$

where $r = |\mathbf{x}| \equiv (\mathbf{x}\cdot\mathbf{x})^{1/2}$. The standard procedure is to represent the exterior field in terms of a set of solutions to the Helmholtz equation in separable coordinates. The spherical and circular representations are

$$f(\mathbf{x}) = \begin{cases} \sum_{n=0}^{\infty} h_n^{(1)}(kr) \sum_{m=-n}^{n} \bar{F}_{nm} Y_{nm}(\theta,\phi), & \text{in 3D,} \\ \sum_{n=-\infty}^{\infty} H_n^{(1)}(kr) \bar{F}_n e^{in\theta}, & \text{in 2D,} \end{cases} \tag{1.3}$$

where the spherical harmonic functions are defined below in (2.6). The representation formulae in (1.3) are complete, and are, in principle, perfectly adequate for source fields measured on the surface of a sphere or circle. However, they are not particularly well suited to directional sources, with a main lobe in a specific direction. Many transducers are explicitly designed to have this property, which is at odds with the representation in

* Department of Mechanical and Aerospace Engineering, Rutgers University, Piscataway, NJ 08855–0909. All thanks to the IMA

† Rome Laboratory ERCT, Hanscom Airforce Base, Hanscom, MA 01731–3010

terms of "global" angular functions, as in equation (1.3). Transducer fields are sometimes well modeled by single or multiple Gaussian beams [1], which suggests that the optimal representation would be in terms of beam-like solutions, rather than spherical harmonic functions.

Our purpose here is to derive alternate representations for the surface and exterior fields in terms of complex point sources. These have the property of being localized about a specific direction, and are closely related to Gaussian beams [2]. The latter are approximate solutions to the Helmholtz equation but are exact solutions to the parabolic equation, which is an asymptotic version of the same equation. Complex point sources are exact solutions to the Helmholtz equations, and the associated representations derived here are exact.

Guassian beams are very popular approximate solutions to the wave equation, as exemplified by the diverse use of the Guassian beams summation technique in different fields *e.g.*, [3]-[6], and see [7] and [8] for reviews. The Gabor representation of data on a line or a plane [9,10] uses Gaussian functions as the basis set, and is an early example of the windowed fourier transform, now widely used. However, the Gabor basis functions can only be propagated as approximate solutions of the wave equation. Complex point sources, on the other hand, have the nice property that they are exact wave fields, and are closely related to Gaussian beams [2], [11]-[13]. They have been used advantageously in modeling complex scattering phenomena that involve Gaussian beam incidence, *e.g.*, [14,15].

Despite their obvious choice for modeling the very useful but simple case of a single Gaussian beam, complex point sources have not been widely used to represent other, directional wave fields. This gap appears strange when one considers that many man-made wave fields are by design very directional. The reason may be ascribed to the lack of an **exact** theory for representing arbitrary source distributions in terms of complex point sources. A suitable representation should be compact in space, like the source field it models, which rules out Gabor-type arrays of infinite extent [9,10]. In an earlier paper [16], one of the authors showed that the simple point source in free space can be replaced by a distribution of complex point sources on a sphere in complex space. In this paper we explore an extension of that approach, to find a way to replace a given radiating wave field by a distribution of complex point sources on a sphere. The complex source/Gaussian beam representations developed here are based upon the representation formulae in equation (1.3), *i.e.*, we will only consider spherical and circular distributions of complex point sources.

2. Introductory equations.

2.1. The pattern function. The wave function $f(\mathbf{x})$ in the exterior region is completely specified by the far-field pattern function F. This is defined by

$$f(\mathbf{x}) = g(\mathbf{x}, 0)\, F(\widehat{\mathbf{x}}), \qquad kr \to \infty, \tag{2.1}$$

where $\widehat{\mathbf{x}} = \mathbf{x}/|\mathbf{x}|$, and $g(\mathbf{x}, \mathbf{y})$ is the free space Green's function, satisfying

$$\nabla^2 g + k^2 g = -\delta(\mathbf{x} - \mathbf{y}). \tag{2.2}$$

Thus,

$$g(\mathbf{x}, \mathbf{y}) = \begin{cases} \frac{i}{4\pi} h_0^{(1)}(k|\mathbf{x} - \mathbf{y}|) = \frac{e^{ik|\mathbf{x}-\mathbf{y}|}}{4\pi|\mathbf{x}-\mathbf{y}|}, & \text{in 3D}, \\ \frac{i}{4} H_0^{(1)}(k|\mathbf{x} - \mathbf{y}|), & \text{in 2D}. \end{cases} \tag{2.3}$$

The pattern function follows from the standard representations in (1.3) as

$$F(\widehat{\mathbf{x}}) = \begin{cases} \sum_{n=0}^{\infty} \sum_{m=-n}^{n} F_{nm} Y_{nm}(\theta, \phi), & \text{in 3D}, \\ \sum_{n=-\infty}^{\infty} F_n e^{in\theta}, & \text{in 2D}, \end{cases} \tag{2.4}$$

where

$$F_{nm} = 4\pi(-i)^{n+1} \bar{F}_{nm}, \qquad F_n = 4(-i)^{n+1} \bar{F}_n, \tag{2.5}$$

and

$$Y_{nm}(\theta, \phi) = \sqrt{\frac{(2n+1)}{4\pi} \frac{(n-m)!}{(n+m)!}}\, P_n^m(\cos\theta) e^{im\phi}. \tag{2.6}$$

The radiated field can be expressed in terms of a Helmholtz integral over a surface S enclosing the source region,

$$f(\mathbf{x}) = \int_S \left\{ f(\mathbf{y}) \frac{\partial g(\mathbf{x}, \mathbf{y})}{\partial n(\mathbf{y})} - g(\mathbf{x}, \mathbf{y}) \frac{\partial f(\mathbf{y})}{\partial n(\mathbf{y})} \right\} \mathrm{d}S(\mathbf{y}), \quad \text{for } \mathbf{x} \text{ in } V, \tag{2.7}$$

where $\mathbf{n}$ is directed into V. The far-field of f follows from equation (2.7) and the far-field expansion of the Green's function,

$$g(\mathbf{x}, \mathbf{y}) = g(\mathbf{x}, 0)\, e^{-ik\widehat{\mathbf{x}}.\mathbf{y}}, \qquad kr \to \infty. \tag{2.8}$$

Thus, from equations (2.1), (2.7), and (2.8),

$$F(\widehat{\mathbf{x}}) = -\int_S \left\{ ik\widehat{\mathbf{x}}.\mathbf{n} f(\mathbf{y}) + \frac{\partial f(\mathbf{y})}{\partial n(\mathbf{y})} \right\} e^{-ik\widehat{\mathbf{x}}.\mathbf{y}} \mathrm{d}S(\mathbf{y}). \tag{2.9}$$

For example, consider initial data on the plane $\mathbf{x}.\boldsymbol{\nu} = 0$ like a Gaussian function centered at the origin:

$$f(\mathbf{x}) = Ae^{-\frac{kr^2}{2a}}, \qquad \frac{\partial f(\mathbf{x})}{\partial n} = ikAe^{-\frac{kr^2}{2a}}, \qquad \text{on } \mathbf{x}.\boldsymbol{\nu} = 0, \tag{2.10}$$

where A in (2.10) is chosen as

$$A = \begin{cases} \frac{i}{4\pi a}, & \text{in 3D,} \\ \frac{i}{(8\pi k a)^{1/2}}, & \text{in 2D,} \end{cases} \tag{2.11}$$

and $\boldsymbol{\nu}$ is a fixed real unit vector. Of course, it is not possible to arbitrarily prescribe both f and $\partial f/\partial n$, as in (2.10), and to simultaneously satisfy the wave equation. The correct procedure calls for one of these to be given, and the other follows from the integral equation which automatically solves the wave equation. However, we will use the pattern function generated by (2.10) for comparison. Thus, substituting (2.10) and (2.11) into equation (2.9) yields

$$F(\widehat{\mathbf{x}}) = \cos^2 \frac{\theta}{2}\, e^{-\frac{1}{2} k a \sin^2 \theta} \quad \text{in 2D and 3D,} \tag{2.12}$$

where θ is the angle between the directions $\widehat{\mathbf{x}}$ and $\boldsymbol{\nu}$, *i.e.*, $\cos\theta = \widehat{\mathbf{x}}.\boldsymbol{\nu}$. The pattern function (2.12) is very similar to that of a complex point source, which is defined next.

2.2. Complex point sources. A complex point source is the analytic extension of the Green's function into complex space. Let the source point be $\mathbf{y} = ia\boldsymbol{\nu}$ where a is real and positive. Define the related wave function which is weighted so that it behaves like a Gaussian beam in a particular direction,

$$G(\mathbf{x}, a\boldsymbol{\nu}) = e^{-ka}\, g(\mathbf{x}, ia\boldsymbol{\nu}). \tag{2.13}$$

It is easily seen that the far-field form of G is simply

$$G(\mathbf{x}, a\boldsymbol{\nu}) = g(\mathbf{x}, 0)\, e^{-ka(1-\widehat{\mathbf{x}}.\boldsymbol{\nu})}, \quad kr \to \infty. \tag{2.14}$$

Let θ be defined as before, *i.e.*, $\cos\theta = \widehat{\mathbf{x}}.\boldsymbol{\nu}$, then

$$G(\mathbf{x}, a\boldsymbol{\nu}) = g(\mathbf{x}, 0)\, e^{-ka2\sin^2\frac{\theta}{2}}, \quad kr \to \infty. \tag{2.15}$$

Its pattern function is therefore

$$F(\widehat{\mathbf{x}}) = \exp\{-ka2\sin^2 \frac{\theta}{2}\}. \tag{2.16}$$

Comparison of this result with the pattern function of equation (2.12) shows that $G(\mathbf{x}, a\boldsymbol{\nu})$ behaves in a manner closely related to the Gaussian initial data of (2.10) and (2.11) on the plane $\mathbf{x}.\boldsymbol{\nu} = 0$. In particular, for large ka, the forward propagated fields, *i.e.*, in $\mathbf{x}.\boldsymbol{\nu} > 0$ are similar, and the fields on $\mathbf{x}.\boldsymbol{\nu} = 0$ match.

The connection between complex point sources and Gaussian beams is normally made in the near field, rather than at infinity. The connection is via the near-axis approximation of the complex length $|\mathbf{x} - ia\boldsymbol{\nu}|$. Let $z = \mathbf{x}.\boldsymbol{\nu}$ and $\rho = |\mathbf{x} \wedge \boldsymbol{\nu}|$, then for $|z - ia| \gg \rho$, or equivalently $z^2 + a^2 \gg \rho^2$,

$$|\mathbf{x} - ia\boldsymbol{\nu}| = z - ia + \frac{\rho^2}{2(z - ia)} + \dots . \tag{2.17}$$

The Gaussian beam approximation then follows from the further assumption that $ka \gg 1$, so that equations (2.3), (2.13), and (2.17), imply to leading order that

$$G(\mathbf{x}, a\boldsymbol{\nu}) \approx e^{ik\left(z + \frac{\rho^2}{2(z-ia)}\right)} \times \begin{cases} \frac{1}{4\pi(z-ia)}, & \text{in 3D,} \\ \frac{e^{i\pi/4}}{\sqrt{8\pi k(z-ia)}}, & \text{in 2D.} \end{cases} \tag{2.18}$$

Note that

$$e^{ik\left(z + \frac{\rho^2}{2(z-ia)}\right)} = e^{\frac{-ka\rho^2}{2(z^2+a^2)}}\, e^{ikz\left(1 + \frac{\rho^2}{2(z^2+a^2)}\right)}, \tag{2.19}$$

and hence the beam width is minimal at $z = 0$. The complex point source is thus closely related to a Gaussian beam both in the near and far-fields. For these reasons, we will refer to the fundamental "complex based Gaussian beam" of equation (2.13) as a CBGB.

3. The beam representation formula. Consider a solution of the Helmholtz equation which is regular outside a sphere of radius b. If $b \geq a$ then it is reasonable to expect that the wave function f can be represented by a distribution of complex point sources placed within the real sphere $r = a$. In particular, we consider a distribution of CBGBs on the complex sphere $\mathbf{x} = ia\boldsymbol{\nu}$ where now $\boldsymbol{\nu}$ is any unit vector on the unit ball. Thus, we try

$$f(\mathbf{x}) = \int_{|\boldsymbol{\nu}|=1} G(\mathbf{x}, a\boldsymbol{\nu}) w(\boldsymbol{\nu}) d\Omega(\boldsymbol{\nu}), \tag{3.1}$$

for some as yet undetermined function w. The far-field pattern of this *ansatz* for f is easily found using equations (2.14) and (3.1). The function f is regular outside $r = b \geq a$, and is therefore completely specified by its far-field pattern. Hence, by comparison with (2.1), it is clear that the weighting function w must satisfy the identity

$$\int_{|\boldsymbol{\nu}|=1} e^{-ka(1-\widehat{\mathbf{x}}.\boldsymbol{\nu})} w(\boldsymbol{\nu}) d\Omega(\boldsymbol{\nu}) = F(\widehat{\mathbf{x}}), \quad \text{for all } |\widehat{\mathbf{x}}| = 1. \tag{3.2}$$

The problem, then, is to find the weighting function w. The 3D and 2D cases are considered separately below.

Norris [16] showed that a point source at the origin, *i.e.*, $g(\mathbf{x}, 0)$, can be represented by an isotropic distribution of point sources on a complex sphere. This is equivalent to the general representation of equation (3.2) for the particular case of $F \equiv 1$. Referring to [16], we see that w = constant in this case, with the explicit results (from equations (4) and (6) of [16]),

$$(3.3)\quad f(\mathbf{x}) = g(\mathbf{x},0) \;\Leftrightarrow\; w = \begin{cases} \frac{e^{ka}}{4\pi j_0(ika)} = \frac{(ka)^{1/2}e^{ka}}{(2\pi)^{3/2}I_{\frac{1}{2}}(ka)}, & \text{in 3D}, \\ \frac{e^{ka}}{2\pi J_0(ika)} = \frac{e^{ka}}{2\pi I_0(ka)}, & \text{in 2D}. \end{cases}$$

Note that the weighting function is purely real in this case.

Before presenting results for w in terms of F, we note that the integral (3.2) can be approximated by

$$(3.4)\qquad F(\widehat{\mathbf{x}}) \sim w(\widehat{\mathbf{x}}) \times \begin{cases} \frac{2\pi}{ka}, & \text{in 3D}, \\ \sqrt{\frac{2\pi}{ka}}, & \text{in 2D}, \end{cases} \qquad ka \to \infty,$$

where we have used the standard asymptotic expressions for a one- and two-dimensional Laplace-type integrals [17, 17, pp. 183, 329]. Thus, for large ka the weighting function w is simply proportional to the far-field pattern.

4. Applications in 2D. Let θ be the polar angle defining $\widehat{\mathbf{x}}$. The far-field pattern $F(\theta)$ for the wave function f can be represented as the Fourier series in (2.4). Similarly, we expand the unknown distribution function as

$$(4.1)\qquad w(\theta) = \sum_{n=-\infty}^{\infty} w_n e^{in\theta}.$$

Substituting into the consistency equation (3.2) yields

$$(4.2)\qquad \sum_{n=-\infty}^{\infty} F_n e^{in\theta} = \sum_{m=-\infty}^{\infty} w_m \int_0^{2\pi} e^{-ka(1-\cos(\psi-\theta))} e^{im\psi}\, \mathrm{d}\psi.$$

A simple change of variables and the orthogonality of the functions $e^{in\theta}$, combined with the identity,

$$(4.3)\qquad \int_0^{2\pi} e^{in\phi + ka\cos\phi} \mathrm{d}\phi = 2\pi\,(-i)^n\, J_n(ika) = 2\pi I_n(ka),$$

implies that

$$(4.4)\qquad w_n = \frac{e^{ka}}{2\pi I_n(ka)} F_n.$$

Note that the ratio w_n/F_n is purely real. The general expressions for the weighting function and the CBGB representation are therefore,

$$(4.5)\qquad w(\theta) = \frac{e^{ka}}{2\pi} \sum_{n=-\infty}^{\infty} \frac{F_n e^{in\theta}}{I_n(ka)},$$

and

$$f(\mathbf{x}) = \frac{e^{ka}}{2\pi} \sum_{n=-\infty}^{\infty} \frac{F_n}{I_n(ka)} \int_0^{2\pi} G(\mathbf{x}, a\boldsymbol{\nu}(\psi)) e^{in\psi} \, d\psi. \tag{4.6}$$

Note that equation (3.3) for the simple, isotropic point source, follows from (4.5) with $F_n = \delta_{n0}$.

4.1. Examples. Let $f = f(r, \theta)$, and let $f_n(r)$ be its fourier coefficients,

$$f(r, \theta) = \sum_{n=-\infty}^{\infty} f_n(r) e^{in\theta} \Leftrightarrow f_n(r) = \bar{F}_n H_n^{(1)}(kr), \tag{4.7}$$

for all $r > r_{source}$ where r_{source} is the radius of the smallest circle enclosing all the sources. Suppose that the radiating wave function $f(\mathbf{x})$ is defined by its value on $r = b$, then

$$f(\mathbf{x}) = \sum_{n=-\infty}^{\infty} f_n(b) \frac{H_n^{(1)}(kr)}{H_n^{(1)}(kb)} e^{in\theta}, \quad r \geq b, \tag{4.8}$$

and the far-field pattern follows from the asymptotic behavior of the Hankel functions for large argument, *i.e.*,

$$F_n = \frac{4(-i)^{n+1}}{H_n^{(1)}(kb)} f_n(b), \tag{4.9}$$

and consequently

$$w(\theta) = \frac{2e^{ka}}{\pi} \sum_{n=-\infty}^{\infty} \frac{(-i)^{n+1} \, e^{in\theta} \, f_n(b)}{I_n(ka) H_n^{(1)}(kb)}. \tag{4.10}$$

As an example, consider a delta function on the circle, $f(b, \theta) = \delta(\theta - \theta_0)$, for which the CBGB weighting function is

$$w(\theta) = \frac{e^{ka}}{\pi^2} \sum_{n=-\infty}^{\infty} \frac{(-i)^{n+1} \, e^{in(\theta-\theta_0)}}{I_n(ka) H_n^{(1)}(kb)}. \tag{4.11}$$

As a second example, consider the Gaussian of equations (2.10) and (2.11) with far-field pattern given by equation (2.12). Using the identity (4.3), it is a simple matter to show that

$$F_n = \begin{cases} \frac{e^{-ka/4}}{2} I_{n/2}(ka/4), & n \text{ even} \\ \frac{e^{-ka/4}}{4} \left[I_{\frac{n-1}{2}}(ka/4) + I_{\frac{n+1}{2}}(ka/4) \right], & n \text{ odd} \end{cases}.$$

and the weighting function follows from equations (4.1) and (4.4). Using equation (4.5), we obtain

$$w(\theta)=\frac{e^{3ka/4}}{4\pi}\sum_{n\text{ even}}\frac{I_{n/2}(ka/4)}{I_n(ka)}e^{in\theta}+\frac{e^{3ka/4}}{8\pi}\sum_{n\text{ odd}}\left[\frac{I_{\frac{n-1}{2}}(ka/4)}{I_n(ka)}+\frac{I_{\frac{n+1}{2}}(ka/4)}{I_n(ka)}\right]e^{in\theta}$$

This is the weighting function required to produce an **exact** radiated field corresponding to the "data" of equation (2.10). We would expect that it closely approximates a delta function in the limit of large ka. Letting $ka \to \infty$ and using the asymptotic formulae for the modified Bessel functions, we find that $w_n \to 1/2\pi$ and hence $w(\theta) \to \delta(\theta)$.

5. Applications in 3D. The direction $\hat{\mathbf{x}}$ is defined by the spherical polar angles θ and ϕ, and the pattern function $F(\theta,\phi)$ is given in equation (2.4). Let the angles θ' and ϕ' define the direction $\boldsymbol{\nu}$, and the CBGB weighting function is

$$w(\theta',\phi') = \sum_{n=0}^{\infty}\sum_{m=-n}^{n} w_{nm}Y_{nm}(\theta',\phi'). \tag{5.1}$$

The consistency equation (3.2) becomes

$$\sum_{n=0}^{\infty}\sum_{m=-n}^{n} F_{nm}Y_{nm}(\theta,\phi) = e^{-ka}\sum_{n=0}^{\infty}\sum_{m=-n}^{n} w_{nm}\int_{4\pi} e^{ka\hat{\mathbf{x}}(\theta,\phi).\boldsymbol{\nu}(\theta',\phi')}Y_{nm}(\theta',\phi')\mathrm{d}\Omega(\theta',\phi'). \tag{5.2}$$

This must be satisfied for all $0 \le \theta \le \pi$ and $0 \le \phi \le 2\pi$.

Start with the identity [18, eq. (11.3.46)]

$$e^{ikr\hat{\mathbf{x}}(\theta,\phi).\boldsymbol{\nu}(\theta',\phi')} = 4\pi\sum_{n=0}^{\infty} i^n j_n(kr)\sum_{m=-n}^{n} Y_{nm}(\theta,\phi)Y^*_{nm}(\theta',\phi'). \tag{5.3}$$

The spherical harmonics are, by definition, orthonormal with respect to integration over the angles, so that

$$\int_{4\pi} e^{ikr\hat{\mathbf{x}}(\theta,\phi).\boldsymbol{\nu}(\theta',\phi')}Y_{nm}(\theta',\phi')\mathrm{d}\Omega(\theta',\phi') = 4\pi i^n j_n(kr)\,Y_{nm}(\theta,\phi). \tag{5.4}$$

Then using the symmetry property $j_n(-z) = (-1)^n j_n(z)$, we find from equations (5.2) and (5.4) that

$$w_{nm} = \frac{i^n e^{ka}}{4\pi j_n(ika)}F_{nm} = \frac{(ka)^{1/2}e^{ka}}{(2\pi)^{3/2}I_{n+\frac{1}{2}}(ka)}F_{nm}. \tag{5.5}$$

It is interesting to note that the ratios w_{nm}/F_{nm} are again purely real.

The general expression for the weighting function follows from equations (5.1) and (5.5) as

$$w(\theta,\phi) = \frac{(ka)^{1/2}e^{ka}}{(2\pi)^{3/2}} \sum_{n=0}^{\infty} \frac{1}{I_{n+\frac{1}{2}}(ka)} \sum_{m=-n}^{n} F_{nm}Y_{nm}(\theta,\phi). \tag{5.6}$$

The special case of a real point source at the origin, which was considered in [16] and is given by equation (3.3), follows immediately from the general result (5.6) with $F_{nm} = \sqrt{4\pi}\,\delta_{n0}\,\delta_{m0}$.

5.1. Examples of prescribed data. Let $f_{nm}(r)$ be the expansion coefficients for the wave function on the sphere $r =$ constant,

$$f(r,\theta,\phi) = \sum_{n=0}^{\infty} \sum_{m=-n}^{n} f_{nm}(r)Y_{nm}(\theta,\phi). \tag{5.7}$$

Consider the reference value $r = b$, then

$$f(\mathbf{x}) = \sum_{n=0}^{\infty} \frac{h_n^{(1)}(kr)}{h_n^{(1)}(kb)} \sum_{m=-n}^{n} f_{nm}(b)Y_{nm}(\theta,\phi), \quad r \geq b, \tag{5.8}$$

and the far-field pattern follows from the asymptotic behavior of the spherical Hankel functions for large argument. The coefficients F_{nm} follow from eqs. $(1.3)_1$, $(2.5)_1$ and (5.8), while (5.5) yields

$$w_{nm} = \left(\frac{2ka}{\pi}\right)^{1/2} \frac{(-i)^{n+1}e^{ka}\,f_{nm}(b)}{I_{n+\frac{1}{2}}(ka)h_n^{(1)}(kb)} \tag{5.9}$$

and hence

$$w(\theta,\phi) = \left(\frac{2ka}{\pi}\right)^{\frac{1}{2}} \sum_{n=0}^{\infty} \frac{(-i)^{n+1}e^{ka}}{I_{n+\frac{1}{2}}(ka)h_n^{(1)}(kb)} \sum_{m=-n}^{n} f_{nm}(b)Y_{nm}(\theta,\phi), \quad a \leq b. \tag{5.10}$$

Using the asymptotic forms of the Bessel functions for large order it is found that the coefficients of (5.9) satisfy

$$|w_{nm}| \sim \text{constant} \cdot \left(\frac{b}{a}\right)^n n\,|f_{nm}(b)|, \qquad \text{as } n \to \infty. \tag{5.11}$$

Thus, the expansion for the weight function $w(\theta,\phi)$ converges slower than the expansion for the field on the sphere $r = b$, eq. (5.7). Also, the fastest convergence of the expansion for w is obtained by choosing $a = b$. The same statement is true for the two-dimensional case.

For example, if the surface function is a delta function,

$$f(b,\theta,\phi) = \frac{\delta(\theta-\theta_0)}{\sin\theta_0}\delta(\phi-\phi_0) = \sum_{n=0}^{\infty}\sum_{m=-n}^{n} Y_{nm}(\theta,\phi)Y^*_{nm}(\theta_0,\phi_0), \tag{5.12}$$

then

$$w(\theta,\phi) = \sum_{n=0}^{\infty} \frac{(-i)^{n+1}e^{ka}\sqrt{2ka/\pi}}{I_{n+\frac{1}{2}}(ka)h_n^{(1)}(kb)} \sum_{m=-n}^{n} Y_{nm}(\theta,\phi)\,Y^*_{nm}(\theta_0,\phi_0). \tag{5.13}$$

6. Convolution representations.

6.1. Inward and outward propagation. Consider the far-field pattern function $F(\theta)$ in 2D. Its fourier coefficients in $(2.4)_2$ imply a multipole expansion at the origin, given by $(1.3)_2$ and $(2.5)_2$. Using $H_n^{(1)}(z) \approx (-i/\pi)(n-1)!(2/z)^n$, for $n \gg |z|$, it follows that the coefficients F_n must decrease pretty quickly in magnitude as $n \to \infty$. Inaccuracies in these coefficients can lead to large errors in the field value for "small" kr. In order to make this more precise, consider the field at two radii: a and b, both greater than r_{source}. The fourier coefficients of $f(a,\theta)$ and $f(b,\theta)$ are related by

$$f_n(a) = \frac{H_n^{(1)}(ka)}{H_n^{(1)}(kb)} f_n(b). \tag{6.1}$$

Write this as

$$f_n(a) = 2\pi f_n(b)Q_n(a,b), \quad \Leftrightarrow \quad f(a,\theta) = f(b,\cdot) * Q(a,b,\cdot)\,(\theta)\,, \tag{6.2}$$

where $*$ denotes convolution,

$$F * c\,(\theta) \equiv \int_0^{2\pi} F(\theta-\theta')c(\theta')d\theta' = \int_0^{2\pi} F(\theta')c(\theta-\theta')d\theta'. \tag{6.3}$$

The Q function is therefore

$$Q(a,b,\theta) = \frac{1}{2\pi}\sum_{n=-\infty}^{\infty} \frac{H_n^{(1)}(ka)}{H_n^{(1)}(kb)} e^{in\theta} = \frac{1}{2\pi}\sum_{n=0}^{\infty} \epsilon_n \frac{H_n^{(1)}(ka)}{H_n^{(1)}(kb)} \cos n\theta, \tag{6.4}$$

where $\epsilon_0 = 1$ and $\epsilon_n = 2$ for $n \geq 1$.

Note that $Q_n(a,b) \approx (2\pi)^{-1}(b/a)^n$ as $n \to \infty$, and the convolution form $(6.2)_2$ of the propagation equation from one radius to another is sensible only if $b \leq a$, that is, the data is propagated from a smaller to a larger radius. The reverse process, back- or inward-propagation, is inherently unstable and cannot be represented as a convolution operation. Propagation from a smaller to a larger radius is a stable process and is simply a convolution operation for the data on the two circles. We note that $Q(a,a,\theta) = \delta(\theta)$ from (6.4), as expected.

The idea of a convolution operation is useful, particularly when the data is spatially restricted to an aperture. This is common in many cases, such as transducers, for which the convolution operation is much more natural than a fourier representation. We will show in the next subsections how this can be used for the CBGB weighting functions. First, we note that the formal definition of the convolution operation $(6.2)_2$ as being equivalent to the relation $(6.2)_1$ between fourier coefficients can be used even when one of the two functions in the convolution operation is undefined. For instance, the function $Q(a,b,\theta)$ with $a < b$ is undefined for all θ because the series in (6.4) is divergent.

6.2. An alternative formula for the 2D weighting function. The formula in (4.5) gives the appropriate weighting of the CBGBs in terms of the fourier components of the far-field, F_n. For practical purposes, the function $F(\theta)$ may be very localized in θ, and a fourier representation is inconvenient. Equation (3.2) is therefore equivalent to $F(\theta) = w * C(\theta)$ with

$$C(\theta) = e^{-ka(1-\cos\theta)}. \tag{6.5}$$

Conversely, (4.5) is the same as $w(\theta) = F * c(\theta)$ where

$$c(\theta) = \frac{e^{ka}}{4\pi^2} \sum_{n=-\infty}^{\infty} \frac{e^{in\theta}}{I_n(ka)} = \frac{e^{ka}}{4\pi^2} \sum_{n=1}^{\infty} \frac{\epsilon_n \cos n\theta}{I_n(ka)}, \tag{6.6}$$

which is again real-valued and symmetric. However, the fact that $I_n(z) \approx (z/2)^n/n!$ as $n \to \infty$ means the coefficients in the sum in (6.6) diverge and the infinite sum is therefore undefined for all θ. However, the function obtained by truncating the sum at any finite n is well defined. The reason why one function, $C(\theta)$ is well defined, and the other, $c(\theta)$, is not is simply a statement of the fact that the integral equation $F(\theta) = w * C(\theta)$ for w does not have an inverse in the form of a convolution operation. Rather, the solution of the integral equation is given by the "weak" form of the convolution, $(6.2)_1$. Finally, we note that (6.5) implies that

$$C(\theta) \to \sqrt{\frac{2\pi}{ka}}\,\delta(\theta) \quad \text{as } ka \to \infty, \tag{6.7}$$

in agreement with the result (3.4).

The propagation formula (6.1) is easily generalized to complex positions. Thus, the field on the complex circle $r = ia$ is defined by

$$Q(ia,b,\theta) = \frac{1}{2\pi} \sum_{n=-\infty}^{+\infty} \frac{H_n^{(1)}(ika)}{H_n^{(1)}(kb)} e^{in\theta} = \frac{1}{\pi^2} \sum_{n=-\infty}^{+\infty} (-i)^{n+1} \frac{K_n(ka)}{H_n^{(1)}(kb)} e^{in\theta}$$

Hence, (6.2), (??) and (4.10) imply that the field on the complex circle is

$$f(ia,\theta) = P(a,\theta) * w(\theta), \tag{6.8}$$

where $P(a,\theta)$ is a real-valued function,

$$P(a,\theta) = \frac{e^{-ka}}{2\pi} \sum_{n=-\infty}^{+\infty} K_n(ka) I_n(ka) e^{in\theta}.$$

It can be easily checked that this sum is convergent for all θ.

Finally, we note that (4.10) is equivalent to

$$w(\theta) = D(a,b,\cdot) * f(b,\cdot), \tag{6.9}$$

where

$$D(a,b,\theta) = \frac{e^{ka}}{\pi^2} \sum_{n=0}^{\infty} \frac{\epsilon_n(-i)^{n+1}\cos n\theta}{I_n(ka)H_n^{(1)}(kb)}. \tag{6.10}$$

This sum converges only if $a > b$.

6.3. A convolution formula for the 3D weighting function. The addition theorem for spherical harmonics [18, p. 1274] is

$$P_n(\cos\omega) = \frac{4\pi}{2n+1} \sum_{m=-n}^{n} Y_{nm}(\theta,\phi) Y^*_{nm}(\theta',\phi'), \tag{6.11}$$

where $0 \le \omega \le \pi$ is the acute angle subtended by the directions (θ,ϕ) and (θ',ϕ'), or $\cos\omega = \cos\theta\cos\theta' + \sin\theta\sin\theta'\cos(\phi-\phi')$. Suppose that the prescribed data is axially symmetric about $\theta = 0$, *i.e.*, $F = F(\theta)$, then $F_{nm} = F_{n0}\delta_{m0}$, and hence, $w_{nm} = w_{n0}\delta_{m0}$ also. Let

$$b(\theta) = \sum_{n=0}^{\infty} \sqrt{\frac{2n+1}{4\pi}}\, b_{n0} Y_{n0}(\theta,\cdot), \tag{6.12}$$

then using (2.4) and the addition formula (6.11), it follows that

$$\int_{4\pi} F(\omega)\, b(\theta') \mathrm{d}\Omega(\theta',\phi') = \sum_{n=0}^{\infty} F_{n0} b_{n0} Y_{n0}(\theta,\cdot). \tag{6.13}$$

Thus, referring to (3.2) and (5.6), we have that

$$F(\theta) = w * C(\theta), \quad w(\theta) = F * c(\theta) \tag{6.14}$$

where $*$ denotes a spherical convolution,

$$w * C(\theta) \equiv \int_{4\pi} w(\omega) C(\theta') \mathrm{d}\Omega(\theta',\phi') = \int_{4\pi} w(\theta') C(\omega) \mathrm{d}\Omega(\theta',\phi'), \tag{6.15}$$

the function $C(\theta)$ is defined in (6.5), and the "function" $c(\theta)$ has fourier coefficients

$$c_{nm} = \frac{(ka)^{1/2} e^{ka}}{4\pi^2} \frac{\sqrt{n+\frac{1}{2}}}{I_{n+\frac{1}{2}}(ka)} \delta_{m0}. \tag{6.16}$$

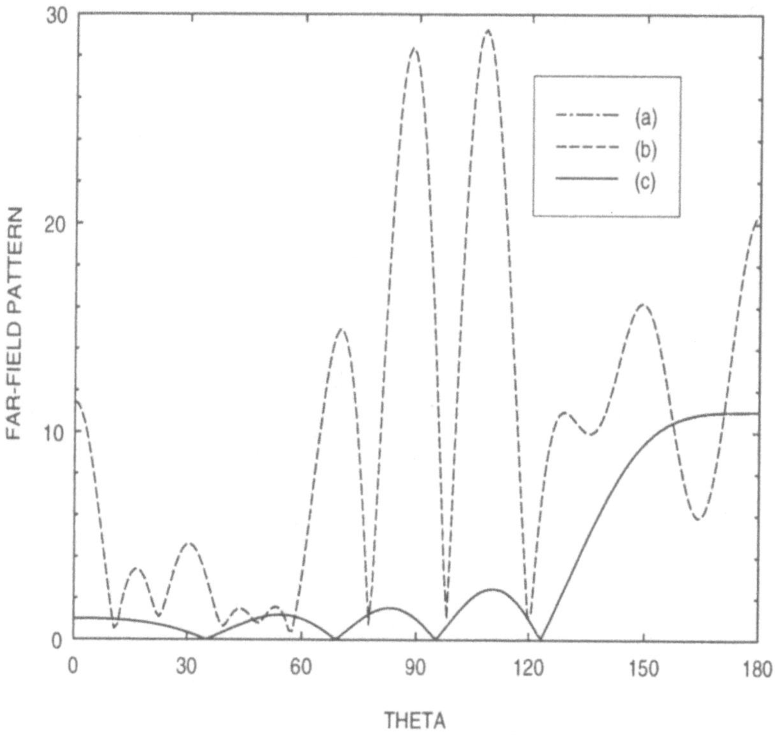

FIG. 7.1. *Far-field pattern of the array containing eleven point sources with* $ka = 10$. *(a) beam formula; (b) Laplace approximation to the beam integral; (c) exact.*

7. Numerical example in 3D. The 3D beam-representation (3.1) will now be verified numerically by computing the far-field pattern of a collection of eleven point sources. Moreover, we shall investigate the accuracy of a simple asymptotic approximation to the beam integral.

The point sources are located on the z axis at $\mathbf{x}_q = (-\lambda + q\Delta z)\,\hat{z}$ where $q = 0, 1, 2, ..., 10$, λ is the wavelength, and $\Delta z = \lambda/5$. There is a phase change of $-k\Delta z$ between two point sources and this change just offsets the propagation-phase advance from point source to point source when the radiation is computed in the negative z direction. This type of array is called an end-fire array and has one main beam directed in the negative z direction. The field of this collection of point sources is given by

$$f(\mathbf{x}) = \sum_{q=0}^{10} g(\mathbf{x}, \mathbf{x}_q)\, e^{-iqk\Delta z}\,, \tag{7.1}$$

where $g(\mathbf{x}, \mathbf{x}_q)$ is the free-space Green's function (2.3). The corresponding

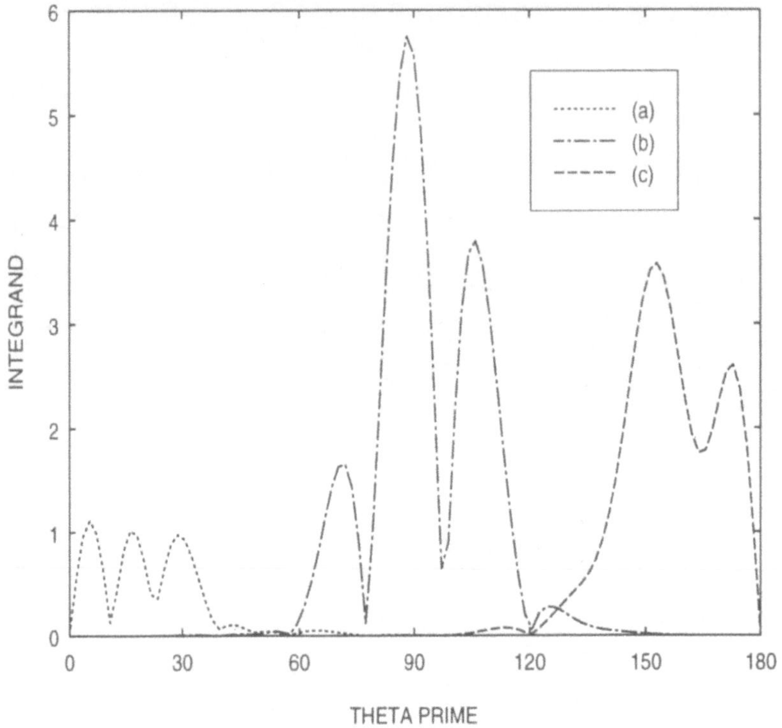

FIG. 7.2. *The integrand of the ϕ-independent beam integral with $ka = 10$ for: (a) $\theta = 0°$; (b) $\theta = 90°$; and (c) $\theta = 180°$.*

far-field pattern is thus a function of θ only:

$$F(\theta, \cdot) = \sum_{q=0}^{10} e^{-ik(-\lambda+q\Delta z)\cos\theta} e^{-iqk\Delta z} \tag{7.2}$$
$$= \left(\frac{1 - e^{-i11k\Delta z(1+\cos\theta)}}{1 - e^{-ik\Delta z(1+\cos\theta)}}\right) e^{-ik\lambda\cos\theta}.$$

To compute the ϕ-independent weighting function $w(\theta, \cdot)$, we shall use the formula (5.10) involving the field on the scan sphere $r = b$. We have $f_{nm}(b) = 0$ for $m \neq 0$ and

$$f_{n0}(b) = 2\pi \int_0^\pi f(b(\sin\theta\hat{x} + \cos\theta\hat{z}))\, Y_{n0}^*(\theta, \cdot) \sin\theta d\theta. \tag{7.3}$$

The number of terms needed to accurately compute the weighting function in the summation (5.10) is determined by the formula $N = [kr_s] + n_1$ where r_s is the radius of the source region and n_1 is a small integer [19, p. 17]. In the present case $r_s = \lambda$ and thus $N = 12$ is sufficiently large. One can then use the formulas of [20, sec 4.1] to compute $f_{n0}(b)$ for $n = 0, 1, 2, ..., 12$ from the values of the field (7.1) on the scan sphere $r = b$ at the angles $\theta = n\Delta\theta$ with $n = 0, 1, 2, ..., 12$, and $\Delta\theta = 2\pi/25$.

In the following numerical calculations we choose the radius of the scan sphere equal to the radius of the sphere that encloses the complex point sources in the beam expansion, that is $b = a$. In this case where the far-field pattern is ϕ independent it is found from (3.2) that

$$(7.4)\quad F(\theta, \cdot) = 2\pi \int_0^\pi e^{-ka(1-\cos\theta\cos\theta')} w(\theta', \cdot) I_0(ka \sin\theta \sin\theta') \sin\theta' d\theta' ,$$

where we have used the integral representation (4.3) for the modified cylindrical Bessel function.

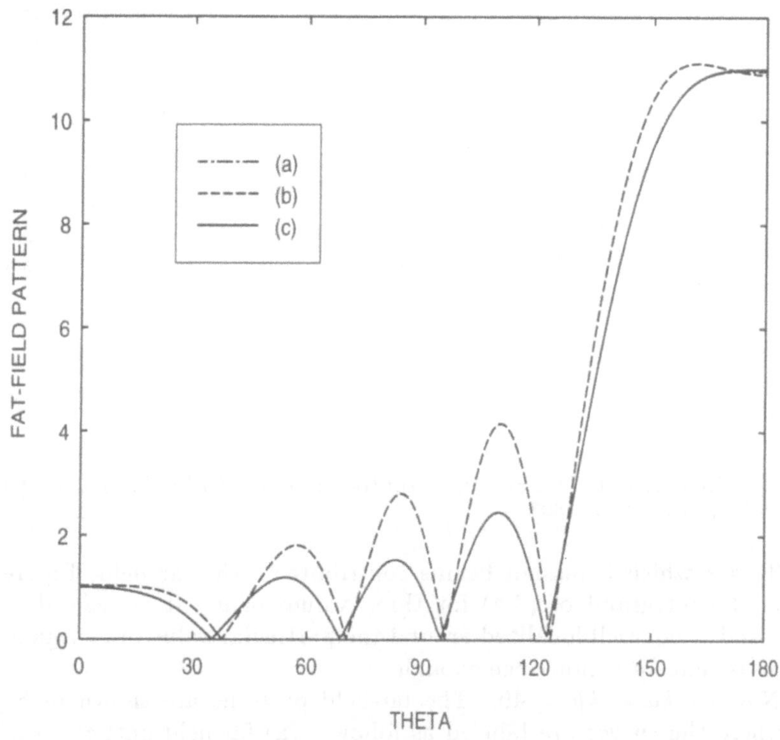

FIG. 7.3. *Far-field pattern of the array containing eleven point sources with ka = 40. (a) beam formula; (b) Laplace approximation to the beam integral; (c) exact.*

Now to the numerical results. Choosing $ka = kb = 10$, the expressions (7.4) and (3.4) give the far-field patterns shown in Figure 7.1, where the curves are labeled as follows: (a) far-field pattern obtained from (7.4); (b) far-field pattern obtained from (3.4); and (c) exact far-field pattern. The result obtained from the beam formula (7.4) is seen to coincide with the exact far-field pattern; whereas the asymptotic result (3.4) does not agree well with the exact far-field pattern. The reason for this discrepancy is, of course, that $ka = 10$ is not large enough for the asymptotic approximation to be valid.

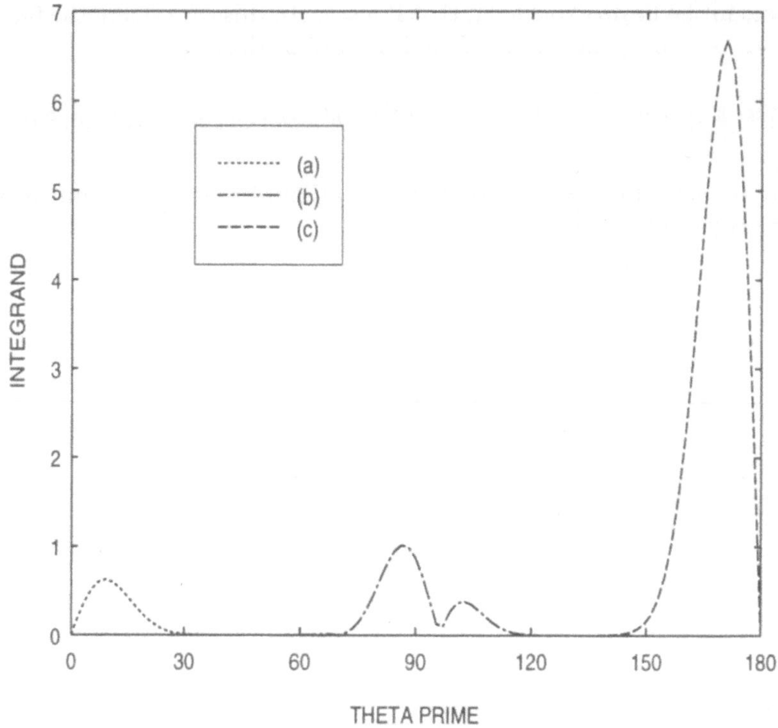

FIG. 7.4. *The integrand of the ϕ-independent beam integral with $ka = 40$ for: (a) $\theta = 0^o$; (b) $\theta = 90^o$; and (c) $\theta = 180^o$.*

To see which Gaussian beams contribute to the far field, Figure 7.2 shows the integrand of (7.4) for three values of θ. It is seen that the integrand is not well localized around the particular value of θ. Again, the reason is that ka is not large enough.

Now let $ka = kb = 40$. The far-field patterns are shown in Figure 7.3 where the curves are labeled as follows: (a) far-field pattern obtained from (7.4); (b) far-field pattern obtained from (3.4); and (c) exact far-field pattern. The result obtained from the beam formula (7.4) is seen to coincide with the exact far-field pattern. Furthermore, the asymptotic result (3.4) agrees fairly well with the exact far-field pattern. Thus, $ka = 40$ is almost large enough for the asymptotic approximation to be valid.

In order to see which Gaussian beams contribute to the far field, Figure 7.4 shows the integrand of (7.4) for three values of θ. It is seen that the integrand is now quite well localized around the particular value of θ.

8. Conclusion. Complex point sources serve as useful building bricks for modeling radiating wave fields. Although they are closely related to Gaussian beams, and can reproduce the radiation pattern of a Gaussian transducer very well, the present results show that they are not restricted to these types of radiation patterns. We have shown that an arbitrary radiation pattern from a compact region can be replicated exactly by a compact distribution of complex point sources. The main results are the formulae for the weighting function w in terms of the radiated field on a sphere or circle in real space, for example, eqs. (4.10) and (5.10) for 2D and 3D, respectively. The weighting function also depends upon the arbitrary value for the distance a representing the complex source radius, which could possibly be used to advantage. This aspect, as well as extensions of the theory to the time domain [21] will be examined elsewhere.

REFERENCES

[1] J. J. Wen and M. Breazeale, *A diffraction beam field expressed as the superposition of Gaussian beams*, J. Acoust. Soc. Am., 83 (1988), pp. 1752–1756.
[2] G. A. Deschamps, *Gaussian beam as a bundle of complex rays*, Electron. Lett. 7 (1971), pp. 684–685.
[3] V. Červený, M. M. Popov, and I. Pšencík, *Computation of wave fields in inhomogeneous media - Gaussian beam approach*, Geophys. J. R. Astron. Soc., 70 (1982), pp. 109–128
[4] M. M. Popov, *A new method of computation of wave fields using Gaussian beams*, Wave Mot., 4 (1982), pp. 85–97.
[5] B. S. White, A. N. Norris, A. Bayliss and R. Burridge, *Some remarks on the Gaussian beam summation method*, Geophys. J. R. Astr. Soc., 89 (1987), pp. 579–636.
[6] M. B. Porter and H. P. Bucker, *Gaussian beam ray tracing for computing ocean acoustic fields*, J. Acoust. Soc. Am., 82 (1987), pp. 1349–1359.
[7] V. Cervený, *Gaussian beam synthetic seismograms*, J. Geophys., 58 (1985), pp. 44–72.
[8] V. M. Babich and M. M. Popov, *Gaussian summation method (review)*, Radiophys. Quant. Electr., 32 (1989), pp. 1063–1081.
[9] M. J. Bastiaans, *Gabor's expansion of a signal into Gaussian elementary signals*, Proc. IEEE, 68 (1980), pp. 538–539.
[10] L. B. Felsen, J. M. Klosner, I. T. Lu, and Z. Grossfeld, *Source field modeling by self-consistent Gaussian beam superposition*, J. Acoust. Soc. Am., 89 (1990), pp. 63–72.
[11] L. B. Felsen, *Complex-point-source solutions of the field equations and their relation to the propagation and scattering of Gaussian beams*, Symp. Math. Instituta di alta Matematica, 18 (1976), pp. 39–56.
[12] M. Couture and P. A. Belanger, *From Gaussian beam to complex-source-point spherical wave*, Phys. Rev. A 24 (1981), pp. 355–359.
[13] L. B. Felsen, *Geometrical theory of diffraction, evanescent waves and complex rays*, Geophys. J. R. Astron. Soc., 79 (1982), pp. 77–88.
[14] H.-C. Choi and J. G. Harris, *Scattering of an ultrasonic beam from a curved interface*, Wave Motion, 11 (1989), pp. 383–406.
[15] S. Zeroug and L. B. Felsen, *Nonspecular reflection of two- and three-dimensional acoustic beams from fluid-immersed plane layered elastic structures*, J. Acoust. Soc. Am., 95 (1994), pp. 3075–3089.
[16] A. N. Norris, *Complex point source representation of real point sources and the*

Gaussian beam summation method, J. Opt. Soc. Am., A3 (1986), pp. 2005–2010.

[17] N. Bleistein and R. A. Handelsman, *Asymptotic Expansions of Integrals*, Dover, New York, 1986.

[18] P. M. Morse and H. Feshbach, *Methods of Theoretical Physics*, New York, McGraw-Hill, 1953.

[19] J. E. Hansen, Ed., J. Hald, F. Jensen, and F. H. Larsen, *Spherical Near-Field Antenna Measurements*, Peter Peregrinus, London, 1988.

[20] T. B. Hansen, *Formulation of spherical near-field scanning in the time domain*, To appear in this IMA Proceedings, 1995.

[21] E. Heyman, *Complex source pulsed beam representation of transient radiation*, Wave Motion, 11 (1989), pp. 337–349.

MODELING SOUND PROPAGATION IN THE OCEAN

MICHAEL B. PORTER*

Abstract. Sound propagates extremely well in the ocean: inexpensive sound projectors and sensors can be used to transmit and receive pulses around the entire globe. Similarly, the rumble and hum of ships acts like an acoustic beacon that allows them to be heard and tracked at long ranges. Still quieter ships can be tracked by shining an acoustic beam on them and looking for the acoustic glint. In civilian applications sound can also be used to image the ocean through ocean acoustic tomography. However, comparing a sound beam to a light beam is at least a little misleading: light is bent under a refractive index but our common experience does little to suggest the severe distortion of a sound beam as it makes its journey through the ocean channel. These effects are usually very important in understanding how acoustic systems function in the ocean. Back-of-the-envelope calculations are too crude and naive numerical approaches too expensive. However, a combination of mathematical and numerical techniques yields elegant tools. We survey these approaches.

1. Introduction. The ocean is a complex swirling body of fluid with large-scale features such as the meandering Gulf Stream and the eddies that are pinched off from it. These features are characterized by their time-varying 3-D profiles of temperature, salinity, and velocity. Their behavior can be modeled by direct solution of the Navier-Stokes equations.

We can perturb this field by submerging a small vibrating object within it. Of course, our background ocean already possesses a cascade of oceanographic features of arbitrary length and temporal scales. However, within the frequency band of human hearing we shall imagine the vibrations of the background ocean as being weak.

The ripples induced by such a vibrating object are also governed by the Navier-Stokes equations. However, a 50 Hz projector generates ripples only 30 m long and it is not computationally practical to solve those equations on such a fine scale over the domains of interest. The usual approach is to treat the ripples as small-amplitude perturbations and linearize the Navier-Stokes equations about the background state. Furthermore, the background is envisioned as a 'frozen ocean' neglecting its time variation. One finds that these perturbations are completely characterized by an acoustic pressure p which satisfies the familiar wave equation:

$$\nabla^2 p - \frac{1}{c^2(\mathbf{x})}\frac{\partial^2 p}{\partial t^2} = -\frac{S(t)}{r}\,\delta(z - z_s, r)\,. \tag{1.1}$$

Here, $c(\mathbf{x})$ is a function of the background salinity, temperature, and pressure and is called the sound speed. The term on the right-hand side models a point-source at a depth z_s vibrating in accordance with a time-series $S(t)$.

* Dept. of Mathematics and Center for Applied Math. and Stats., New Jersey Institute of Technology, Newark, NJ 07102. Work supported in part by ONR contracts N00014-92-J and N00014-95-1-0558.

Note that the total pressure in the ocean is then obtained by adding the pressure of these small (acoustic) perturbations to the background pressure. Thus, small ripples ride on the backs of the larger ripples.

Computational ocean acoustics is mostly concerned with solving Eq. 1.1 while ocean circulation modeling is concerned with the modeling of large-scale features using the original Navier-Stokes equations (or other simplifications of those equations). The material parameters (such as the sound speed) that go into the equation are generally measured in situ.

The most obvious approach to numerically treating the wave equation in Eq. 1.1 is to directly discretize it using finite-difference or finite-element methods. Interestingly, this is seldom practical in ocean acoustics problems because they often involve 100's of wavelengths in depth and 1000's of wavelengths in range. A variety of mathematical techniques must be invoked first in order to reduce the wave equation to a tractable form.

The first such simplification is the assumption that the source emits a pure tone ($S(t) = e^{i\omega t}$). This yields the familiar Helmholtz (or reduced wave equation):

$$\nabla^2 p + \frac{\omega^2}{c^2(\mathbf{x})} p = -\frac{1}{r}\,\delta(z - z_s, r)\,. \tag{1.2}$$

Of course, if we can solve the Helmholtz equation for an arbitrary frequency we can then compute the solution for an arbitrary waveform by just summing across its component frequencies. However, sources of interest are actually often tonals.

Unfortunately, even the Helmholtz equation remains too complicated for direct numerical treatment and further simplifications are needed. The following mathematical techniques are useful: high-frequency asymptotics, separation of variables, and wave factorization. These ideas form the basis of ray, normal mode, spectral integral, and parabolic equation methods. (The interrelations between the mathematical methods is an interesting discussion itself[1].)

In the remaining sections we review the mathematical derivation of each of these approaches. We then discuss the practical computational difficulties of the methods and the open research issues. Lastly we show how each of the model types is used on typical ocean acoustics problems. The reader interested in a detailed discussion of the numerics is referred to Ref. [2].

2. Rays. To derive the ray equations one seeks a solution of the Helmholtz equation in the form:

$$p(r,z) = e^{ik\phi(r,z)} \sum_{j=0}^{\infty} A_j(r,z) \frac{1}{(ik)^j}\,, \tag{2.1}$$

where $k = \omega/c_0$ is a wavenumber defined with respect to an arbitrary wave speed c_0 (which might be the mean speed of sound in the ocean). This

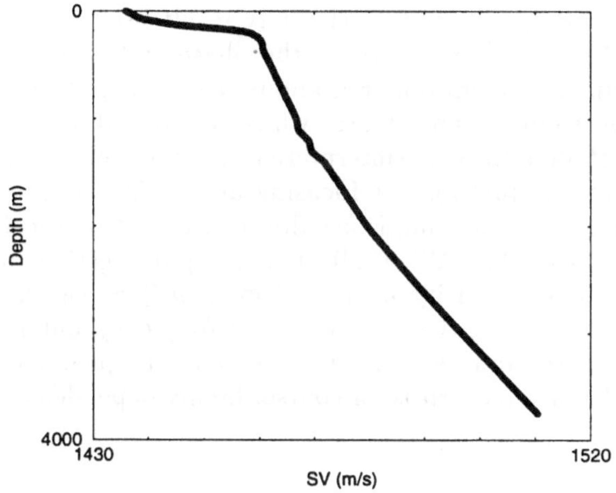

FIG. 2.1. *Sound speed profile for the Arctic case.*

substitution converts the original Helmholtz equation (a linear PDE) into an infinite sequence of nonlinear PDE's whose solution gives the phase $\phi(r,z)$ and the amplitudes $A_j(r,z)$:

$$\begin{aligned}(\nabla\phi)^2 &= -c_0^2/c^2(r,z),\\ 2\nabla\phi\cdot\nabla A_0+(\Delta\phi)A_0 &= 0,\\ 2\nabla\phi\cdot\nabla A_j+(\Delta\phi)A_j &= -\Delta A_{j-1}, \qquad j=1,2,\ldots.\end{aligned} \tag{2.2}$$

So far there appears to be little gain. However, for high frequencies (ω large and therefore k large) one may be content to keep the leading term (A_0) in this asymptotic series. Furthermore, the equations for the phase and amplitude are easily solved by introducing a family of curves (the rays) which satisfy a simple initial value problem:

$$\begin{aligned}\frac{dr}{ds} &= c\rho(s), & \frac{d\rho}{ds} &= -\frac{1}{c^2}\frac{dc}{dr},\\ \frac{dz}{ds} &= c\zeta(s), & \frac{d\zeta}{ds} &= -\frac{1}{c^2}\frac{dc}{dz},\end{aligned} \tag{2.3}$$

with $(r(0),z(0),\rho(0),\zeta(0)) = (0,z_s,\cos\alpha/c(0),\sin\alpha/c(0))$. These initial conditions imply that the ray is launched from the source position and with a take-off angle of α. The solution of the IVP gives the ray trajectory, $(r(s),z(s))$ where s is arclength along the ray.

To trace the rays, we need a sound-speed profile such as the example in Fig. 2.1 taken from the Arctic. Then, solving the ray equations with a sequence of different take-off angles α we construct a ray fan as shown in Fig. 2.2.

The ray trace is really just an intermediate step in calculating the acoustic field; however, it is often the only result used from a ray model. To complete the process one rewrites the phase and amplitude equations in a new coordinate system centered about the ray trajectories. Along the rays the phase is simply an integral while the amplitude is governed by the spreading of adjacent rays (interpreted in a differential sense). The eye is quite good at interpreting this focusing and de-focusing in a ray trace. Computationally the spreading is usually measured numerically by simple finite-difference formulas. When all of this is put together one obtains a plot of the acoustic intensity in dB ($10 \log_{10} |p|^2$) as shown in Fig. 2.3. The ray paths are insensitive to the source frequency but not the phase along each ray path. Here we have used a source frequency of 50 Hz. The resulting interference pattern is, of course, highly dependent on frequency.

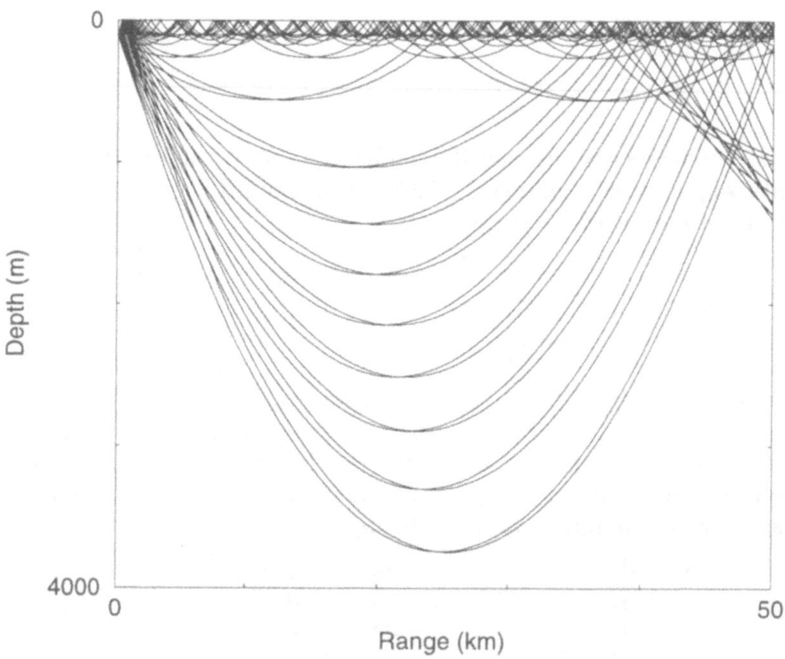

FIG. 2.2. *Ray trace for the Arctic case.*

We have described a process in which (1) high-frequency asymptotics is exploited to produce a simple IVP, (2) the IVP is solved using standard numerical solvers, and (3) an intensity field is calculated. This follows a standard mathematical presentation but it overlooks difficulties which are so severe that ray methods have some disrepute in SONAR modeling. Some

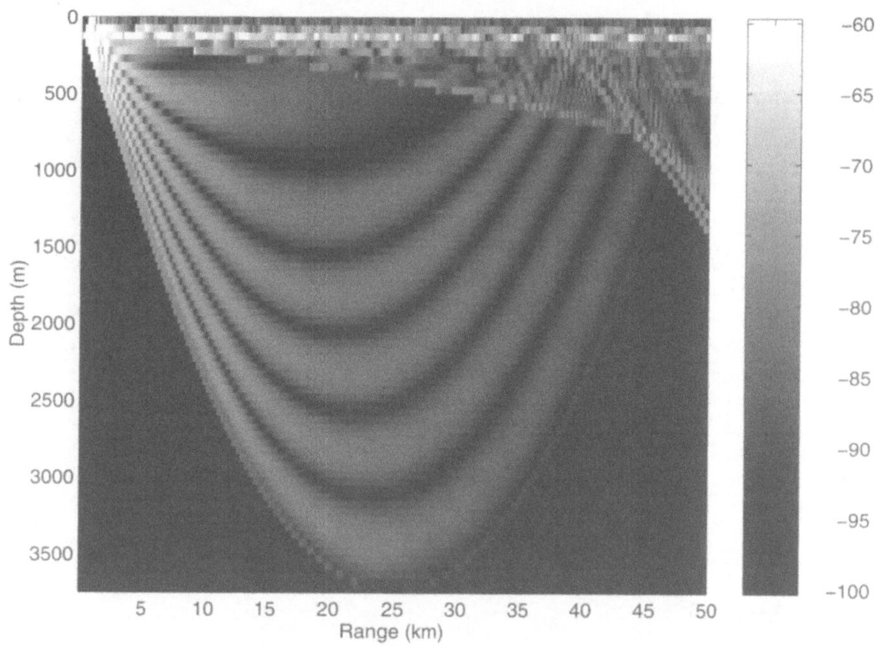

FIG. 2.3. *Intensity for the Arctic case.*

of these difficulties are much less severe for stratified problems. However, normal mode or spectral integral approaches are usually far better for such problems. Thus, we must have in mind a range-dependent scenario.

Amongst the practical difficulties of ray models is the problem of finding *eigenrays*. These are defined as all those rays that connect the source to a fixed receiver position. The acoustic field at any given point is gotten by summing up the contributions from each of the eigenrays. However, finding the eigenrays requires solving a nonlinear, and only piecewise-smooth equation for the take-off angle α that launches a ray to the desired receiver position. This process must be repeated thousands of times to build up a complete picture of the acoustic field.

The difficulty of finding eigenrays is further complicated when one includes ray paths that penetrate into the ocean bottom. The ocean bottom is typically characterized by interfaces at which an incoming ray is split between transmitted and reflected rays. As the process is repeated, one ray leads to a cascade of rays which must be accounted for. It is also noteworthy that the ray equations are a dynamical system that is quite commonly chaotic[3].

A further difficulty is that the ray approximation is usually non-uniform: the predicted intensity goes to infinity at focal regions or caustics. Of course, uniform corrections can be used but the numerical problem of

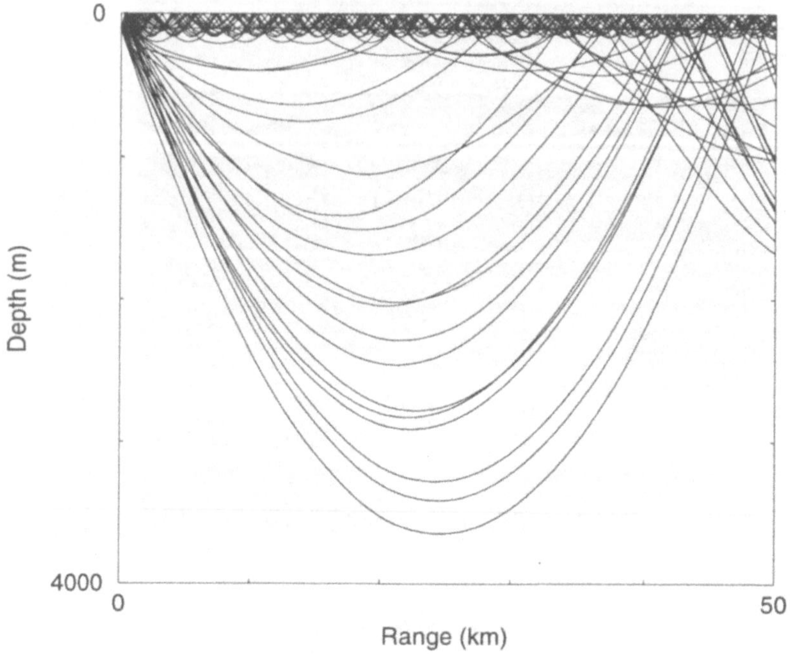

FIG. 2.4. *Ray trace for the Arctic case using an unfiltered sound speed profile.*

identifying the caustic locations is non-trivial. These zones where the ray approximation yields a poor approximation may sometimes be tolerated. However, caustics also cause phase changes along the rays. We must keep track of these phase changes (to form the so-called KMAH index) even if we do not intend to apply a uniform correction. This again requires identifying the locations of caustics.

The Arctic sound speed profile shown in Fig. 2.1 has small scale features. At low frequencies such features are unseen in the sense that the acoustic field is not sensitive to them. However, ray theory in essence tries to extrapolate the high-frequency behavior (where such features are important) down to lower frequencies. If we had not smoothed the sound speed profile in Fig. 2.1 before doing the ray trace we would have obtained a ragged picture as shown in Fig. 2.4. The irregular behavior of the rays is in turn echoed in the resulting acoustic field (which becomes quite inaccurate). While this filtering solves the problem, there is no formal theory that prescribes how this should optimally be done.

The sound speed profile is usually provided at discrete depths rather than as an analytic function. Because of the sensitivity to the profile interpolation, the overall numerical accuracy depends as much on the interpolation as on the control of truncation error in the numerical integrator.

Considering all the above effects one typically finds that ray models are erratic, with their reliability depending a great deal on the insight and experience of the user.

3. Spectral integral and modal representations. The key approximation in spectral integral and modal representations is that of stratification, i.e. that the sound speed depends only on depth and not range. (However, solutions of the stratified problem can be used to approximate range-dependent problems by gluing together a sequence of range-independent segments.) When the problem is stratified, a Hankel transform eliminates the range variable from Eq. 1.2. The transformed pressure,

$$G(k,z) = \int_0^\infty p(r,z)\, J_0(kr)\, r\, dr, \tag{3.1}$$

then satisfies the following BVP:

$$\begin{aligned} \frac{d^2G}{dz^2} + \left(\frac{\omega^2}{c^2(z)} - k^2\right) G &= \delta(z - z_s), \\ G(0) = 0, \qquad \frac{dG}{dz}(D) &= 0. \end{aligned} \tag{3.2}$$

The boundary conditions correspond to a pressure release surface and a rigid bottom.

The numerical procedure is then to solve Eq. 3.2 for a sequence of wavenumbers $k_l = l\,\delta k$. This can be done using standard finite-difference or finite-element methods though coefficient approximation has normally been favored. Then, the inverse transform

$$p(r,z) = \int_0^\infty G(k,z)\, J_0(kr)\, k\, dk \tag{3.3}$$

can be evaluated approximately as a sum of panels (rectangle or trapezoidal rule):

$$p(r,z) \approx \sum_l G(k_l,z)\, J_0(k_l r)\, k_l\, \delta k. \tag{3.4}$$

When the field is needed for many ranges, an FFT is usually the preferred way to calculate the integral. Usually $G(k)$ is negligible for $k > k_{max}$ where

$$k_{max} = \max_z \frac{\omega}{c(z)},$$

so that the inverse transform is well-approximated over a finite domain of wavenumbers.

For large ranges the kernel in the inverse transform oscillates very rapidly and therefore must be sampled at many wavenumbers. The usual way of coping with this is to evaluate the integral by residues. Note that the Green's function can be written

$$g(z) = \frac{1}{2\pi\rho(z_s)} \sum_m \frac{\Psi_m(z_s)\,\Psi_m(z)}{k^2 - k_m^2}, \tag{3.5}$$

where Ψ_m are normalized eigenfunctions of the homogeneous Sturm-Liouville problem:

$$\frac{d^2\Psi_m(z)}{dz^2} + \left[\frac{\omega^2}{c^2(z)} - k_m^2\right]\Psi_m(z) = 0, \tag{3.6}$$

with

$$\Psi(0) = 0, \qquad \left.\frac{d\Psi}{dz}\right|_{z=D} = 0. \tag{3.7}$$

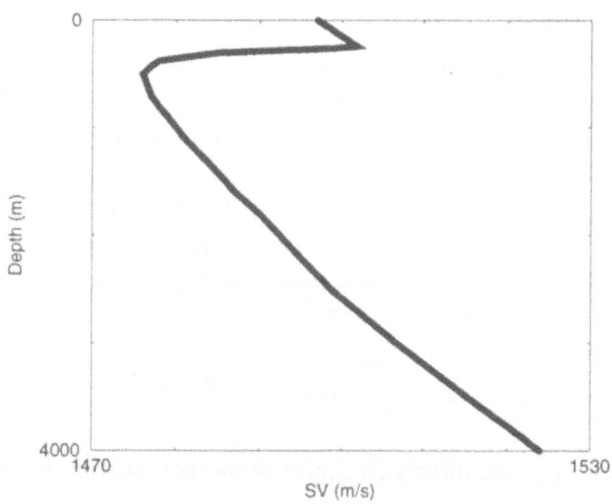

FIG. 3.1. *Sound speed profile for the surface duct problem.*

With this representation of the Green's function it is easy to evaluate the residues and calculate the contour integral. The result is:

$$p(r, z) = \frac{i}{4} \sum_{m=1}^{\infty} \Psi_j(z_s)\,\Psi_j(z)\,H_0^{(1)}(k_j r), \tag{3.8}$$

where k_j are the poles. This of course is the same result one would obtain by separation of variables.

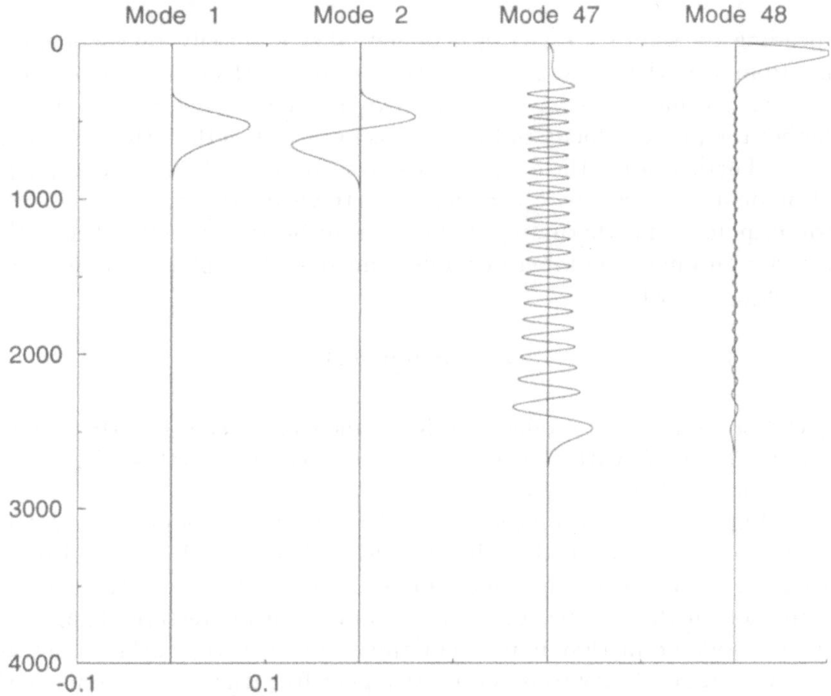

FIG. 3.2. *Selected modes for the surface duct problem.*

The primary numerical problem is now that of finding the eigensolution of the Sturm-Liouville problem. Again, finite-difference or finite-element methods provide a viable solution. However, there are many practical difficulties which we shall discuss in the context of an example.

Wind-driven mixing at the ocean surface usually leads to the formation of an surface duct as seen in the upper 250 m in Fig. 3.1. The ocean then acts like a double-well potential with some modes trapped in the surface duct and some trapped in the main duct. To understand this more clearly it is useful to rewrite the modal equation as:

$$\frac{d^2\Psi_m(z)}{dz^2} + \omega^2 \left[\frac{1}{c^2(z)} - \frac{1}{c_m^2}\right] \Psi_m(z) = 0\,, \tag{3.9}$$

where $c_m = \omega/k_m$, i.e. the phase velocity of the mth mode. Now we can see that the ODE assumes the forms:

$$\frac{d^2\Psi_m(z)}{dz^2} + \gamma^2 \Psi_m = 0 \tag{3.10}$$

$$\frac{d^2\Psi_m(z)}{dz^2} - \gamma^2 \Psi_m = 0, \tag{3.11}$$

depending on whether $c_m > c(z)$ or not. Let us briefly review some results from WKBJ theory. At any particular depth the eigenfunctions have either an oscillatory behavior or an exponential behavior depending on whether the phase velocity of the mode is more or less than the local sound speed. Furthermore, the behavior changes from oscillatory to exponential at depths, z, where $c_m = c(z)$ and these depths are referred to as turning points. Furthermore, we know from Sturm-Liouville theory that there are an infinite number of modes and that their phase speeds satisfy $c_m \in [c_{min}, \infty)$ where

$$c_{min} = \min_z c(z).$$

We choose to label the modes in order of their phase speed so that mode 1 refers to the mode with the slowest phase speed. (With this labeling, the mth mode also has m zeros.)

In Fig. 3.2 we have plotted selected modes for a frequency of 120 Hz. Modes 1 and 2 are trapped in the main SOFAR (Sound Fixing and RAnging) duct. As we go up in mode number to Mode 47 we see the expected progression in the number of zeros. However, an interesting change happens to mode 48 in that it is essentially trapped in the surface duct and bears a great similarity to mode 1. At higher frequencies the modes in the surface duct are even more decoupled from those in the main duct.

Numerically, such multiple-duct problems require careful treatment. If they are solved by shooting methods then the IVP's involve integration through unstable zones (where both growing and decaying exponentials are present). Multiple shooting[5] or re-orthogonalization is needed. Similarly, methods based on coefficient approximation must be carefully constructed so that important decaying exponential solutions are not lost in the discretization noise of a finite-precision calculation.

Finite-difference methods yield algebraic eigenvalue problems (AEVPs) which can be solved by QR or spectrum-slicing methods. This approach is essentially free of stability problems because such problems have been already carefully addressed in the well-developed numerical methods for AEVPs.

Once the modes have been calculated, they can be summed to produce a plot of the acoustic intensity as shown in Fig. 3.3. Here we have placed the source at 25 m and therefore in the surface duct. From plots of the modes and tables of the eigenvalues, it is hard for the human mind to visualize how the modes sum up in and out of phase to reveal this complicated acoustic field. Ray theory provides the language to discuss the field: rays with a shallow take-off angle are trapped in the surface duct producing a band of energy in that same duct. As the rays go to steeper angles significant energy is carried through complex rays into the SOFAR duct (tunneling) where the ray trajectories again become real. (This is seen as the band of energy that leaks out of the surface duct.) Still steeper rays form the

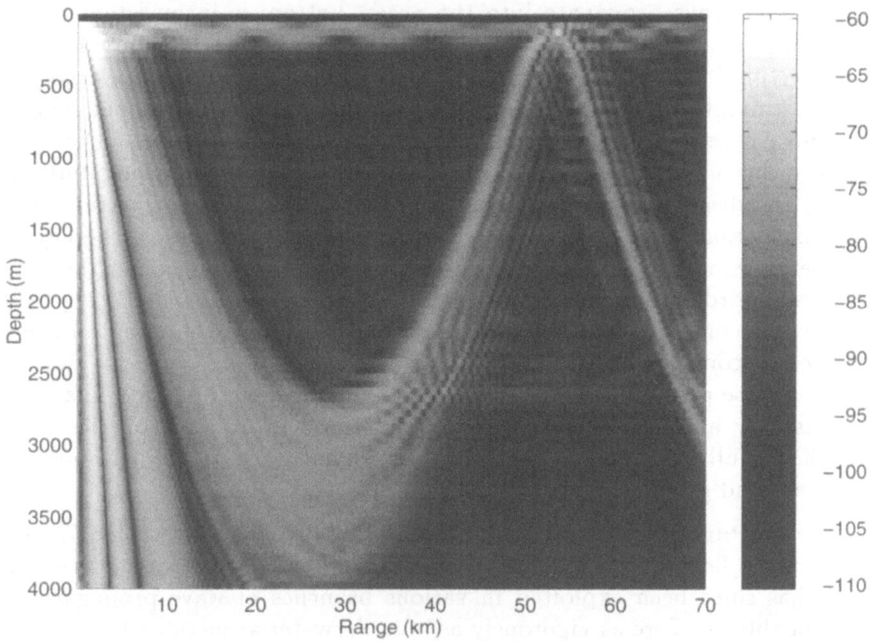

FIG. 3.3. *Intensity for the surface duct problem.*

so-called convergence zone pattern involving the deep-cycling band that comes back to the surface duct at a range of about 50 km. Finally, the steepest rays strike the bottom and are absorbed.

This surface duct example illustrates many interesting features which are discussed more thoroughly in Ref. [7]. To our knowledge, a complete mathematical analysis has never been done. However, broadband acoustic calculations have been done to observe how simple pulses propagate. In addition, analytical studies of wave propagation in the simple case of one linear layer overlying another motivate the complex or leaky ray interpretation. It is found that the rays that leak out of the surface duct into the SOFAR duct are refracted back to the surface. Furthermore, the leaky rays are all phase locked in the sense that their trajectories form a family of translated ray paths each with the same phase delay. Because of this phase-locking the leaky paths can add up destructively across a long stretch in the surface duct and blank out the acoustic field there. Whether the leaky rays add up constructively or destructively is an extremely sensitive function of the details of the sound-speed profile: a change of just 1 m/s in the surface duct is enough to cause 15 dB effects on the intensity throughout the surface duct. Lastly, because of this sensitivity many acoustic models fail on this problem.

The astute reader will have noted that our comment that the steepest angle rays propagate into the ocean bottom is inconsistent with the rigid boundary condition described above (which would imply a perfectly reflecting bottom). In fact, model users in ocean acoustics are accustomed to terminating the problem at some depth with a homogeneous halfspace. This causes serious complications in the numerical models: (1) the spectrum becomes mixed with both discrete and continuous contributions[6], (2) the discrete spectrum involves trapped and leaky (or virtual) modes which *grow* exponentially with depth: the modal series is therefore not everywhere convergent, (3) the resulting AEVP no longer fits the pattern of standard solvers found, for instance, in EISPACK, (4) the eigenvalues become complex and therefore the root search must be done on a multi-sheeted complex plane.

These problems become all the more challenging when attenuation and elasticity are introduced as is now common in ocean acoustics models. Then, multiple roots are possible[4] as are backward traveling modes (with phase and group velocity having opposite sign).

4. Parabolic equations. The parabolic equation was originally derived by Leontovich and Fock[8] in the context of electromagnetic waves. It has since been exploited in various branches of wave propagation but probably nowhere as vigorously as in underwater acoustics where it was introduced by Tappert and Hardin[9,10]. (Interestingly, it is only in the last few years that the PE method has again found significant interest in the radio-wave propagation problems whence it originated.) We shall first outline the derivation by considering the far simpler case of waves on a string. We pick a reference sound speed c_0 and then define the index of refraction $n(z) = c_0/c(z)$ and the wavenumber $k_0 = \omega/c_0$. The one-dimensional version of the Helmholtz equation is then:

$$\frac{d^2\psi}{dx^2} + k_0^2 n^2(x)\psi = 0. \tag{4.1}$$

This equation should be coupled with a boundary condition at $x = 0$ representing the source and a radiation condition as $x \to \infty$. We observe that if the index of refraction is constant, then the general solution is

$$\psi(x) = Ae^{ik_0x} + Be^{-ik_0x}, \tag{4.2}$$

and the particular solution that satisfies the radiation condition is:

$$\psi(x) = e^{ik_0x}. \tag{4.3}$$

Numerically, the problem could be solved by finite-differences which would produce a tridiagonal system of equations. The objective of the parabolic equation approach is to convert this global BVP to a local IVP that can simply be marched out in range. To do this, we write our equation in operator form as

$$P^2 + Q^2\psi = 0, \tag{4.4}$$

where,

$$P = \frac{d}{dx}, \quad Q = n^2. \tag{4.5}$$

We can then factor the equation:

$$(P - ik_0Q)(P + ik_0Q)\psi - ik_0[P, Q]\psi = 0\,, \tag{4.6}$$

where $[P, Q]$ is the so-called *commutator* that measures the degree to which the operators P and Q fail to commute:

$$[P, Q]\psi = PQ\psi - QP\psi. \tag{4.7}$$

The commutator vanishes when P and Q commute, for instance, when the index of refraction does not vary with x. The standard argument is then that if the range-variation is gradual, the commutator is negligible. Furthermore, there will be negligible backscatter and we can consider the part that governs only the right-traveling wave:

$$(P - ik_0Q)\psi = 0\,, \tag{4.8}$$

i.e.,

$$\frac{d\psi}{dx} + ik_0n(x)\psi = 0\,. \tag{4.9}$$

This equation is now first-order and is properly posed with just the single initial condition representing the source. Thus we have accomplished our goal or converting the BVP to a simpler, marchable IVP. We observe that if the index of refraction is constant, we recover the exact solution given in Eq. 4.3.

Now let us repeat the derivation in the more interesting PDE case. The Helmholtz equation then takes the form:

$$\left[P^2 + 2ik_0P + k_0^2(Q^2 - 1)\right]\psi = 0, \tag{4.10}$$

where,

$$P = \frac{\partial}{\partial r}, \quad Q = \sqrt{n^2 + \frac{1}{k_0^2}\frac{\partial^2}{\partial z^2}}. \tag{4.11}$$

We factor this equation as

$$(P + ik_0 - ik_0Q)(P + ik_0 + ik_0Q)\psi - ik_0[P, Q]\psi = 0, \tag{4.12}$$

where $[P, Q]$ is a commutator that again vanishes in range-independent media. Neglecting the commutator and retaining only the factor that corresponds to outgoing waves we obtain:

$$P\psi = ik_0(Q - 1)\psi, \tag{4.13}$$

or,

$$\frac{\partial \psi}{\partial r} = ik_0 \left(\sqrt{n^2 + \frac{1}{k_0^2} \frac{\partial^2}{\partial z^2}} - 1 \right) \psi. \tag{4.14}$$

Thus, we have converted the original elliptic Helmholtz equation to an equation that is first-order in the range-derivative and solvable as an IVP like the heat equation. Here we have avoided calling it parabolic for it is technically pseudo-differential and we must further explain how we intend to interpret the operator on the right-hand side with its roots of derivatives. Usually this is done by using a Taylor series expansion of the square root. Thus one writes:

$$Q = \sqrt{1+q}, \tag{4.15}$$

where, $q = \varepsilon + \mu$ and

$$\varepsilon = n^2 - 1, \qquad \mu = \frac{1}{k_0^2} \frac{\partial^2}{\partial z^2}. \tag{4.16}$$

The small q approximation to the square root is $\sqrt{1+q} = 1 + \frac{q}{2}$. (One can show that the small q approximation equates to narrow angle propagation.) Putting this all together one obtains the standard parabolic equation:

$$\frac{\partial \psi}{\partial r} = \frac{ik_0}{2} \left(n^2 - 1 + \frac{1}{k_0^2} \frac{\partial^2}{\partial z^2} \right) \psi. \tag{4.17}$$

It should be obvious that there are some delicate steps in the above derivation. Consider that the solution of the 1-D outgoing equation is

$$\psi(x) = e^{ik_0 \int_0^x n(\xi)\, d\xi}, \tag{4.18}$$

while the WKB approximation to the solution of the 1-D Helmholtz equation is:

$$\psi(x) \approx \frac{e^{ik_0 \int_0^x n(\xi)\, d\xi}}{\sqrt{n(x)}}. \tag{4.19}$$

Thus, the process of factorizing and discarding the commutator has corrupted the amplitude of the solution. This is discussed more completely in Ref. [13].

For problems with range-dependence, a PE solution is often the method of choice. As an example, we consider a deep-water case with a sound speed profile as shown in Fig. 4.1. The range-dependent feature introduced in this case is an idealized seamount having the form of an isosceles triangle. It is centered at a range of 100 km and extends from 80 – 120 km. Its peak rises 2000 m above the ocean bottom.

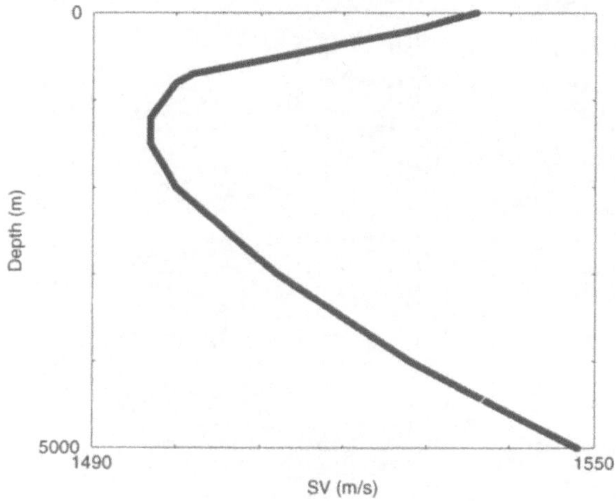

FIG. 4.1. *Sound speed profile for the seamount problem.*

In Fig. 4.2 we show the acoustic intensity due to a 50 Hz source located at a depth of 100 m. In the near field we see a typical dipole pattern arising from the interference of the source with its reflection in the ocean surface. This gives rise to a multilobe beam pattern. The beams then propagate towards the ocean bottom where they split into reflected and transmitted beams. However, the beam with the shallowest take-off angle is refracted before hitting the bottom and cycles back towards the ocean surface. In the absence of the seamount it would continue to cycle up and down the water column with very little energy loss. (Since it is not hitting the lossy bottom boundary the main mechanism for intensity decay is simply cylindrical spreading.) However, at about 90 km the refracted beam hits the seamount which disturbs its cyclical pattern.

In recent years, elastic and poro-elastic PE models have been developed for ocean acoustic problems. In addition, codes have been written to handle an arbitrary number of terms in the Taylor expansion of the square-root operator[11] using a Padé series formulation due to Bamberger[12]. This yields models that are accurate for very steep angle propagation. Nevertheless, there are still some important unanswered questions. We have seen how a careless factorization can yield models that violate energy conservation. These issues have not been thoroughly addressed for elastic and poro-elastic PE models. Furthermore, most of the models require the user to perform a convergence study with respect to range and depth grids. Work is needed to produce robust schemes to automatically control the error.

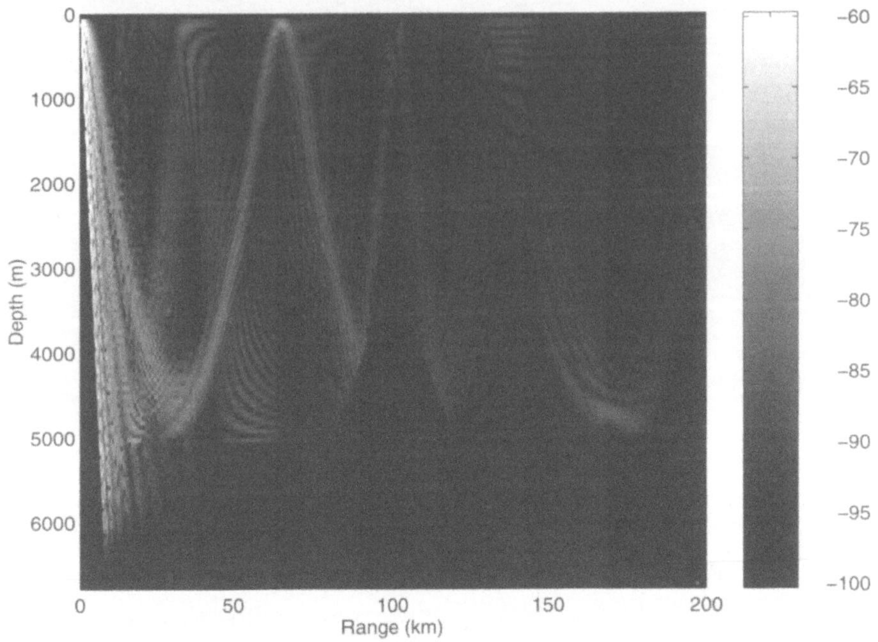

FIG. 4.2. *Intensity for the seamount problem.*

5. Summary. Ocean acoustics has largely been driven by passive and active SONAR applications. New applications coupled with improved computer performance have led to the development of sophisticated modeling tools that combine interesting numerical and mathematical techniques. Of course, this process continues today although civilian applications have become more important than in the past. For instance, sound is being considered as a tool for monitoring global warming. The ocean sound speed increases with temperature so that one can take the ocean's temperature by measuring the time-of-flight of an acoustic pulse. In one scenario a set of acoustic sources would be distributed around the world. A network of receivers would then be used to measure ocean temperature along a set of paths criss-crossing the world's oceans.

Does such a scheme make sense? To further study that issue an experiment called the Heard Island Feasibility Test (HIFT) was conducted in 1991[14]. A source was deployed near Heard Island in the Indian Ocean and received by both ship-towed arrays and fixed surveillance arrays operated by a host of international participants. Acoustic models such as those described above were extremely important in predicting where a ship should sail to hear the sounds. After the experiment the models played an equally important role in interpreting the results.

This experiment led to many interesting new areas of research primar-

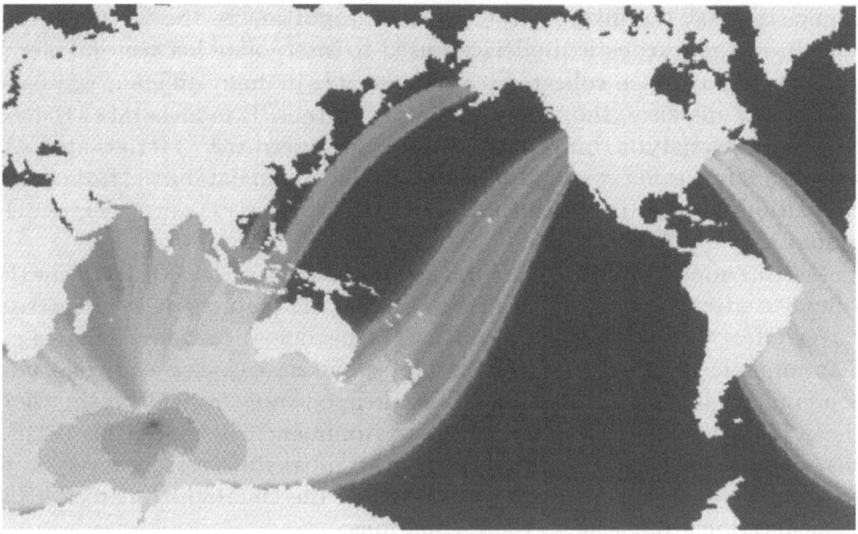

FIG. 5.1. *Intensity due to a* 1 Hz *source near Heard Island (from Collins, et al.[15].*

ily because of the extremely long propagation paths. First, the models had mostly been written for a cylindrical coordinate system as described above; it was clearly necessary to account for the earth curvature effects in HIFT. Furthermore, the deviations from sphericity also turned out to be important. Lastly, the numerical reliability of various models over such enormous lengths is an issue still requiring study.

A glimpse of the interesting tools that have emerged from HIFT is provided in Fig. 5.1 which shows a prediction of acoustic intensity due to the source at Heard Island. This particular result was produced using a 3D PE model[15]. The curved paths look somewhat artificial on this flattened projection but they are actually quite close to great circle paths on the globe. One can see the shadowing due to bathymetric features such as continents and islands. (The band of energy that seems to penetrate the U.S. is actually two beams which have traveled in opposite directions around the globe to either coast of the U.S.)

Global acoustics is just one new area of development in underwater sound. Mine-hunting sonar has generated strong interest in high-frequency acoustics; for a variety of reasons active sonar systems have become of greater interest; and changes in global politics have shifted the emphasis from deep-water to shallow water environments. With these shifts, new and interesting problems in ocean acoustics have emerged.

We have here presented computational ocean acoustics as the problem of solving the Helmholtz equation in two and three dimensions. One cannot go much further in a brief introduction to the subject. However, we should emphasize that the inverse problem is as important as the forward problem. In essence, acoustic models are used to interpolate between measured data which have been collected by many people in many different ways and places. For instance, the whole derivation of the PE models takes the existence of an analytic, range-dependent sound speed $c(r, z)$ for granted. At sea, such a function is pieced together from historical databases, from ocean circulation models, from instruments that sample $c(r, z)$ in an extremely limited way.

Furthermore, bubble clouds near the ocean surface and fish within the volume scatter the sound. In general, fine scale features such as internal waves affect the acoustic propagation. One does not imagine having a deterministic knowledge of this sort of environment; however, one may contemplate stochastic modeling. Even then, one must consider from where the statistical characterization of the environment will come. The PDE's that arise in ocean acoustics are interesting in themselves; however, as in so many fields, the hard part is not to solve the equations but to ask the right questions that lead to those equations.

REFERENCES

[1] D.S. Ahluwalia and J.B. Keller, *"Exact and asymptotic representations of the sound field in a stratified ocean"*, in *Wave Propagation and Underwater Acoustics*, edited by J.B. Keller and J.S. Papadakis (Springer-Verlag, New York, 1977), pp. 14–85.

[2] F. Jensen, W. Kuperman, M. Porter and H. Schmidt, *Computational Ocean Acoustics*, American Institute of Physics, New York (1994).

[3] K.B. Smith, M.G. Brown, and F.D. Tappert, *"Ray chaos in underwater acoustics"*, J. Acoust. Soc. Am. **91**, 1939-1949 (1992).

[4] R.B. Evans, *"The existence of generalized eigenfunctions and multiple eigenvalues in underwater acoustics"*, J. Acoust. Soc. Am. **92**, 2024–2029 (1992).

[5] H.B. Keller, *Numerical solution of two point boundary value problems*, (SIAM, Philadelphia, 1976).

[6] D.C. Stickler, *"Normal-mode program with both the discrete and branch line contributions"*, J. Acoust. Soc. Am. **57**, 856–861 (1970).

[7] Michael B. Porter and Finn B. Jensen, *"Anomalous PE results for propagation in leaky surface ducts"*, J. Acoust. Soc. Am. **94**(3):1510–1516 (1993).

[8] M.A. Leontovich and V.A. Fock, Zh. Eksp. Teor. Fiz. **16**, 557–573 (1946) [Engl. transl.: J. Phys. USSR **10**, 13–24 (1946)].

[9] R.H. Hardin and F.D. Tappert, *"Applications of the split-step Fourier method to the numerical solution of nonlinear and variable coefficient wave*, equations," SIAM Rev. **15**, 423 (1973).

[10] F.D. Tappert, *"The parabolic approximation method"*, in *Wave Propagation in Underwater Acoustics*, edited by J.B. Keller and J.S. Papadakis (Springer-Verlag, New York, 1977), pp. 224–287.

[11] M.D. Collins, *"A higher-order parabolic equation for wave propagation in an ocean overlying an elastic bottom"*, J. Acoust. Soc. Am. **86**, 1459–1464 (1989).

[12] A. Bamberger, B. Engquist, L. Halpern, and P. Joly, *"Higher order parabolic wave equation approximations in heterogeneous media"*, SIAM J. Appl. Math. **48**, 129–154 (1988).

[13] M. B. Porter, Carlo M. Ferla and F. Jensen, *"The Problem of Energy Conservation in One-Way Equations"*, J. Acoust. Soc. Am. **89**(3):1058–1067 (1991).

[14] Walter Munk and Arthur Baggeroer, *The Heard Island papers: A contribution to global acoustics*", J. Acoust. Soc. Am. **96**(4):2327–2329 (1994).

[15] Michael D. Collins, B. Edward McDonald, Kevin D. Heaney and W.A. Kuperman, "*Three-dimensional effects in global acoustics*", J. Acoust. Soc. Am. **97**(3):1567–1575 (1995).

IMA SUMMER PROGRAMS

1987 Robotics
1988 Signal Processing
1989 Robustness, Diagnostics, Computing and Graphics in Statistics
1990 Radar and Sonar (June 18 - June 29)
New Directions in Time Series Analysis (July 2 - July 27)
1991 Semiconductors
1992 Environmental Studies: Mathematical, Computational, and Statistical Analysis
1993 Modeling, Mesh Generation, and Adaptive Numerical Methods for Partial Differential Equations
1994 Molecular Biology
1995 Large Scale Optimizations with Applications to Inverse Problems, Optimal Control and Design, and Molecular and Structural Optimization
1996 Emerging Applications of Number Theory
1997 Statistics in Health Sciences.

SPRINGER LECTURE NOTES FROM THE IMA:

The Mathematics and Physics of Disordered Media
Editors: Barry Hughes and Barry Ninham
(Lecture Notes in Math., Volume 1035, 1983)

Orienting Polymers
Editor: J.L. Ericksen
(Lecture Notes in Math., Volume 1063, 1984)

New Perspectives in Thermodynamics
Editor: James Serrin
(Springer-Verlag, 1986)

Models of Economic Dynamics
Editor: Hugo Sonnenschein
(Lecture Notes in Econ., Volume 264, 1986)

The IMA Volumes in Mathematics and its Applications

Current Volumes:

1 **Homogenization and Effective Moduli of Materials and Media** J. Ericksen, D. Kinderlehrer, R. Kohn, and J.-L. Lions (eds.)

2 **Oscillation Theory, Computation, and Methods of Compensated Compactness** C. Dafermos, J. Ericksen, D. Kinderlehrer, and M. Slemrod (eds.)

3 **Metastability and Incompletely Posed Problems** S. Antman, J. Ericksen, D. Kinderlehrer, and I. Muller (eds.)

4 **Dynamical Problems in Continuum Physics** J. Bona, C. Dafermos, J. Ericksen, and D. Kinderlehrer (eds.)

5 **Theory and Applications of Liquid Crystals** J. Ericksen and D. Kinderlehrer (eds.)

6 **Amorphous Polymers and Non-Newtonian Fluids** C. Dafermos, J. Ericksen, and D. Kinderlehrer (eds.)

7 **Random Media** G. Papanicolaou (ed.)

8 **Percolation Theory and Ergodic Theory of Infinite Particle Systems** H. Kesten (ed.)

9 **Hydrodynamic Behavior and Interacting Particle Systems** G. Papanicolaou (ed.)

10 **Stochastic Differential Systems, Stochastic Control Theory, and Applications** W. Fleming and P.-L. Lions (eds.)

11 **Numerical Simulation in Oil Recovery** M.F. Wheeler (ed.)

12 **Computational Fluid Dynamics and Reacting Gas Flows** B. Engquist, M. Luskin, and A. Majda (eds.)

13 **Numerical Algorithms for Parallel Computer Architectures** M.H. Schultz (ed.)

14 **Mathematical Aspects of Scientific Software** J.R. Rice (ed.)

15 **Mathematical Frontiers in Computational Chemical Physics** D. Truhlar (ed.)

16 **Mathematics in Industrial Problems** A. Friedman

17 **Applications of Combinatorics and Graph Theory to the Biological and Social Sciences** F. Roberts (ed.)

18 ***q*-Series and Partitions** D. Stanton (ed.)

19 **Invariant Theory and Tableaux** D. Stanton (ed.)

20 **Coding Theory and Design Theory Part I: Coding Theory** D. Ray-Chaudhuri (ed.)

21 **Coding Theory and Design Theory Part II: Design Theory** D. Ray-Chaudhuri (ed.)

22 **Signal Processing Part I: Signal Processing Theory** L. Auslander, F.A. Grünbaum, J.W. Helton, T. Kailath, P. Khargonekar, and S. Mitter (eds.)

23 **Signal Processing Part II: Control Theory and Applications of Signal Processing** L. Auslander, F.A. Grünbaum, J.W. Helton, T. Kailath, P. Khargonekar, and S. Mitter (eds.)
24 **Mathematics in Industrial Problems, Part 2** A. Friedman
25 **Solitons in Physics, Mathematics, and Nonlinear Optics** P.J. Olver and D.H. Sattinger (eds.)
26 **Two Phase Flows and Waves** D.D. Joseph and D.G. Schaeffer (eds.)
27 **Nonlinear Evolution Equations that Change Type** B.L. Keyfitz and M. Shearer (eds.)
28 **Computer Aided Proofs in Analysis** K. Meyer and D. Schmidt (eds.)
29 **Multidimensional Hyperbolic Problems and Computations** A. Majda and J. Glimm (eds.)
30 **Microlocal Analysis and Nonlinear Waves** M. Beals, R. Melrose, and J. Rauch (eds.)
31 **Mathematics in Industrial Problems, Part 3** A. Friedman
32 **Radar and Sonar, Part I** R. Blahut, W. Miller, Jr., and C. Wilcox
33 **Directions in Robust Statistics and Diagnostics: Part I** W.A. Stahel and S. Weisberg (eds.)
34 **Directions in Robust Statistics and Diagnostics: Part II** W.A. Stahel and S. Weisberg (eds.)
35 **Dynamical Issues in Combustion Theory** P. Fife, A. Liñán, and F.A. Williams (eds.)
36 **Computing and Graphics in Statistics** A. Buja and P. Tukey (eds.)
37 **Patterns and Dynamics in Reactive Media** H. Swinney, G. Aris, and D. Aronson (eds.)
38 **Mathematics in Industrial Problems, Part 4** A. Friedman
39 **Radar and Sonar, Part II** F.A. Grünbaum, M. Bernfeld, and R.E. Blahut (eds.)
40 **Nonlinear Phenomena in Atmospheric and Oceanic Sciences** G.F. Carnevale and R.T. Pierrehumbert (eds.)
41 **Chaotic Processes in the Geological Sciences** D.A. Yuen (ed.)
42 **Partial Differential Equations with Minimal Smoothness and Applications** B. Dahlberg, E. Fabes, R. Fefferman, D. Jerison, C. Kenig, and J. Pipher (eds.)
43 **On the Evolution of Phase Boundaries** M.E. Gurtin and G.B. McFadden
44 **Twist Mappings and Their Applications** R. McGehee and K.R. Meyer (eds.)
45 **New Directions in Time Series Analysis, Part I** D. Brillinger, P. Caines, J. Geweke, E. Parzen, M. Rosenblatt, and M.S. Taqqu (eds.)

46 **New Directions in Time Series Analysis, Part II**
D. Brillinger, P. Caines, J. Geweke, E. Parzen, M. Rosenblatt, and M.S. Taqqu (eds.)

47 **Degenerate Diffusions**
W.-M. Ni, L.A. Peletier, and J.-L. Vazquez (eds.)

48 **Linear Algebra, Markov Chains, and Queueing Models**
C.D. Meyer and R.J. Plemmons (eds.)

49 **Mathematics in Industrial Problems, Part 5** A. Friedman

50 **Combinatorial and Graph-Theoretic Problems in Linear Algebra**
R.A. Brualdi, S. Friedland, and V. Klee (eds.)

51 **Statistical Thermodynamics and Differential Geometry of Microstructured Materials**
H.T. Davis and J.C.C. Nitsche (eds.)

52 **Shock Induced Transitions and Phase Structures in General Media** J.E. Dunn, R. Fosdick, and M. Slemrod (eds.)

53 **Variational and Free Boundary Problems**
A. Friedman and J. Spruck (eds.)

54 **Microstructure and Phase Transitions**
D. Kinderlehrer, R. James, M. Luskin, and J.L. Ericksen (eds.)

55 **Turbulence in Fluid Flows: A Dynamical Systems Approach**
G.R. Sell, C. Foias, and R. Temam (eds.)

56 **Graph Theory and Sparse Matrix Computation**
A. George, J.R. Gilbert, and J.W.H. Liu (eds.)

57 **Mathematics in Industrial Problems, Part 6** A. Friedman

58 **Semiconductors, Part I**
W.M. Coughran, Jr., J. Cole, P. Lloyd, and J. White (eds.)

59 **Semiconductors, Part II**
W.M. Coughran, Jr., J. Cole, P. Lloyd, and J. White (eds.)

60 **Recent Advances in Iterative Methods**
G. Golub, A. Greenbaum, and M. Luskin (eds.)

61 **Free Boundaries in Viscous Flows**
R.A. Brown and S.H. Davis (eds.)

62 **Linear Algebra for Control Theory**
P. Van Dooren and B. Wyman (eds.)

63 **Hamiltonian Dynamical Systems: History, Theory, and Applications**
H.S. Dumas, K.R. Meyer, and D.S. Schmidt (eds.)

64 **Systems and Control Theory for Power Systems**
J.H. Chow, P.V. Kokotovic, R.J. Thomas (eds.)

65 **Mathematical Finance**
M.H.A. Davis, D. Duffie, W.H. Fleming, and S.E. Shreve (eds.)

66 **Robust Control Theory** B.A. Francis and P.P. Khargonekar (eds.)

67 **Mathematics in Industrial Problems, Part 7** A. Friedman

68 **Flow Control** M.D. Gunzburger (ed.)

69 **Linear Algebra for Signal Processing**
A. Bojanczyk and G. Cybenko (eds.)
70 **Control and Optimal Design of Distributed Parameter Systems**
J.E. Lagnese, D.L. Russell, and L.W. White (eds.)
71 **Stochastic Networks** F.P. Kelly and R.J. Williams (eds.)
72 **Discrete Probability and Algorithms**
D. Aldous, P. Diaconis, J. Spencer, and J.M. Steele (eds.)
73 **Discrete Event Systems, Manufacturing Systems, and Communication Networks**
P.R. Kumar and P.P. Varaiya (eds.)
74 **Adaptive Control, Filtering, and Signal Processing**
K.J. Åström, G.C. Goodwin, and P.R. Kumar (eds.)
75 **Modeling, Mesh Generation, and Adaptive Numerical Methods for Partial Differential Equations** I. Babuska, J.E. Flaherty, W.D. Henshaw, J.E. Hopcroft, J.E. Oliger, and T. Tezduyar (eds.)
76 **Random Discrete Structures** D. Aldous and R. Pemantle (eds.)
77 **Nonlinear Stochastic PDEs: Hydrodynamic Limit and Burgers' Turbulence** T. Funaki and W.A. Woyczynski (eds.)
78 **Nonsmooth Analysis and Geometric Methods in Deterministic Optimal Control** B.S. Mordukhovich and H.J. Sussmann (eds.)
79 **Environmental Studies: Mathematical, Computational, and Statistical Analysis** M.F. Wheeler (ed.)
80 **Image Models (and their Speech Model Cousins)**
S.E. Levinson and L. Shepp (eds.)
81 **Genetic Mapping and DNA Sequencing**
T. Speed and M.S. Waterman (eds.)
82 **Mathematical Approaches to Biomolecular Structure and Dynamics**
J.P. Mesirov, K. Schulten, and D. Sumners (eds.)
83 **Mathematics in Industrial Problems, Part 8** A. Friedman
84 **Classical and Modern Branching Processes**
K.B. Athreya and P. Jagers (eds.)
85 **Stochastic Models in Geosystems**
S.A. Molchanov and W.A. Woyczynski (eds.)
86 **Computational Wave Propagation**
B. Engquist and G.A. Kriegsmann (eds.)

FORTHCOMING VOLUMES

1993–1994: *Emerging Applications of Probability*
Mathematical Population Genetics
1994–1995: *Waves and Scattering*
Wavelet, Multigrid and Other Fast Algorithms (Multiple, FFT) and Their Use in Wave Propagation
Waves in Random and Other Complex Media
Inverse Problems in Wave Propagation
Singularities and Oscillations
Quasiclassical Methods
Multiparticle Quantum Scattering with Applications to Nuclear, Atomic, and Molecular Physics
1995 Summer Program: *Large Scale Optimization with Applications to Inverse Problems, Optimal Control and Design, and Molecular and Structural Optimization*
1995–1996: *Mathematical Methods in Materials Science*
Mechanical Response of Materials from Angstroms to Meters
Phase Transformations, Composite Materials, and Microstructure
Disordered Materials
Particulate Flows: Processing and Rheology
Interface and Thin Films
Nonlinear Optical Materials
Numerical Methods for Polymeric Systems
Topology and Geometry in Polymer Science
Mathematics in Industrial Problems, Part 9
1996 Summer Program: *Emerging Applications of Number Theory*
Applications and Theory of Random Sets